Media Reform and the Climate Emergency

Media Reform and the Climate Emergency

Rethinking Communication in the Struggle for a Sustainable Future

DAVID J. PARK

UNIVERSITY OF MICHIGAN PRESS
ANN ARBOR

First paperback edition 2025

Published in the United States of America by the
University of Michigan Press
First published in paperback July 2025

A CIP catalog record for this book is available from the British Library.

Library of Congress Cataloging-in-Publication Data

Names: Park, David J., author.
Title: Media reform and the climate emergency : rethinking communication in the struggle for a sustainable future / David J. Park.
Description: Ann Arbor : University of Michigan Press, 2021. | Includes bibliographical references and index. |
Identifiers: LCCN 2021009769 (print) | LCCN 2021009770 (ebook) | ISBN 9780472132713 (hardcover : alk. paper) | ISBN 9780472129331 (ebook)
Subjects: LCSH: Climatic changes in mass media. | Mass media and the environment—United States. | Climate change mitigation—United States.
Classification: LCC P96.C58 P37 2021 (print) | LCC P96.C58 (ebook) | DDC 320.58—dc23
LC record available at https://lccn.loc.gov/2021009769
LC ebook record available at https://lccn.loc.gov/2021009770

ISBN 978-0-472-04003-2 (pbk.)

Cover illustration: Julia Hawkins (CC-BY 2.0) and Pixabay / Pete Linforth, adapted by Valentine Seebart.

Authorized Representative: Easy Access System Europe, Mustamäe tee 50, 10621 Tallinn, Estonia, gpsr.requests@easproject.com

Acknowledgments

While it appears at the end of 2020 that Joe Biden is likely to become the next president of the United States, it remains to be seen whether his policies will be bold enough to adequately address the climate emergency through extensive regulation, or if he will support watered-down policies that avoid addressing the root causes of the climate crisis. Time will soon tell us how the fossil fuel interests will interact with Biden's team. Most news media are reporting positively on Biden's very racially diverse proposed cabinet, and both environmentalists *and* the fossil fuel industry are praising his nominations. Critical analyses of the effectiveness of his future climate policies, however, are essentially nonexistent in the mainstream media. While the early praise from the fossil fuel industry is most likely image-driven, it may also be sincere. Early support from one of the most powerful, deceitful, and environmentally harmful industries in the world is concerning.

News outlets are becoming increasingly partisan, and many mainstream news outlets seem supportive of the Democratic Party's conservative wing, so future studies will need to examine whether and how critically news media scrutinize Biden's climate policies. Beyond being attentive to media analyses, citizens will also need to be more active and organized in the fight for a sustainable planet and future. It is extremely likely we will all need to exert significant pressure on a Biden Presidency if we want immediate climate action.

This book was written with a sense of urgency, given the increasing momentum of the climate crisis and the immediate need to find new and creative ways to help mitigate the emergency; I wanted to offer suggestions, but I don't purport to have the final word on the subject. A couple of the chapters constitute significantly revised versions of previously published articles. Many scholars and anonymous reviewers have contributed their thoughts, critiques, and revisions at various stages of this book's development. In particular, I want to thank James Hamilton, Robert Bodle, and Ezequiel Korin, editors of *Explorations in Critical Studies of Advertising*, and Wayne Hope at the *Political Economy of Communication* journal for helping me develop earlier ideas, arguments, and areas of focus. I would also like to thank the editors and affiliated scholars/reviewers at the *International Journal of Social Ecology and Sustainable Development* and at *Integrated Environmental Assessment and Management.* My friends at *Ámbitos: Revista Internacional de Comunicación* in Seville, Spain also deserve appreciation.

I am especially thankful to Elizabeth Demers, Editorial Director at the University of Michigan Press, for her patience, guidance, and wisdom with this book. I would also like to thank John Raymond for his fantastic copyediting and impressive eye for detail, which greatly improved the overall flow and content presentation. The anonymous reviewers of the book manuscript also offered immensely productive feedback, making this a much stronger and cohesive book. Michigan's Editorial Assistant Haley Winkle was also very helpful, kind, and patient with my questions, while everyone else at the Press who helped make this and all of their other books possible deserve recognition as well.

I am also grateful to my family and ancestors whose choices in life were able to influence the good things in mine. Thank you, Mom and Dad, Charles and Edna Park, for your belief in me, for your love and constant support. The same goes for my sister, Laura, who is a wonderfully talented filmmaker and photographer. I am eternally grateful to Leslie Richardson, without whom much of this book would not have been possible. Her intellectual challenges, advice, long discussions, and support were pivotal to the successful completion of this book. We are all individuals made up of experiences and influences from people we encounter and learn from. Please forgive me if I got something wrong in this book. Any errors should be solely ascribable to me. Last, I want to thank the many scholars, activists, politicians, and concerned citizens who continue to fight for a more sustainable, just, and equal world.

Contents

Digital materials related to this title can be found on the Fulcrum platform via the following citable URL: https://doi.org/10.3998/mpub.11927458

Introduction

> A society in which consumption has to be artificially stimulated in order to keep production going is a society founded on trash and waste, and such a society is a house built upon sand.
> —Dorothy L. Sayers[1]

This book is about how specific aspects of mass communication interact with and contribute to the global climate emergency.[2] It asks whether and how mass communication can help address and mitigate the crisis, especially when there is a diminishing window of time to lower emissions that contribute to climate change. At this stage, there should be no doubt that what humanity is facing is both a *crisis* and *emergency*, and both terms are used interchangeably throughout this book. *Climate change* and *global warming* are also often used interchangeably, even though climate change refers more broadly to transformations in various climate patterns and global warming refers to increases in the planet's temperature. This reflects the acceptance of their interchangeability within popular and policy discourse.[3] Indeed, global warming, which exists as part of an overall process of climate change, is already happening and if we don't stop its human-made causes quickly, scientists predict that the climate will spiral out of human control, causing massive planetary and lifeform damage.

Given the gravity of the emergency, many scholars and scientists are working overtime to present solutions. Scientists in agriculture, land use, biology, sea-level rise, and other areas are pitching in, but contributions regarding mitigation from the mass communication field appear limited. This is surprising. Mass communication is extremely interdisciplinary, as it often forces researchers to grapple with multiple fields and conceptual frameworks at the same time in order to address complex problems.[4] This is a strength, and in this sense, the field is one of the few that is uniquely positioned to connect many of the forces that influence the global climate emergency. Economic systems, advertising, governments, approaches to production and consumption, news media, politics, globalization, the military, and more all impact the crisis, and mass communication is critical in each of these domains.

This book outlines the historical contexts of a number of these forces, showing how they have shaped our present challenges, and offers practical solutions to help address a terrifying problem. It places immense significance on the creation of a consumer society, how this society influenced mass communication, and ultimately how both have influenced the climate crisis. Capitalism eventually created an industrial social organization that produces an immense quantity of goods and services for consumption. Much of the world's population lives in these consumer societies where economies are driven by the consumption of seemingly limitless products and services. Many infrastructures, including mass communication infrastructures and processes, are largely planned around this social organization, often by fulfilling its needs while extending demand for its way of life. The challenge with this organization is that the production and consumption of goods and services in consumer societies are central to the creation of greenhouse gas (GHG) emissions, which significantly contribute to the global warming crisis.[5] Individual consumption in consumer-oriented economic systems accounts for the majority of emissions, while indirect emissions generated from the production processes of goods and services account for the highest percentage of harmful emissions.[6] Consumer societies fuel human-made global warming, and much of mass communication is either dependent upon or closely tied to the success of this social organization.

This means that much of advertising, the digital infrastructure, and what's left of journalism, as part of our mass communication system, are pushing humanity in the wrong direction. All encourage overconsumption or are economically supported by commercial organizations that profit from overconsumption. This means advertising, digital media, and much of journalism encourage, perpetuate, and help extend the climate crisis. Our mass communication system is not structured to help move humanity out of this emergency because too many of its goals, objectives, and funding mechanisms clash with what's needed to effectively address the crisis. Our challenge is to reduce overall production and consumption to adequately address the climate emergency, but much of our mass communication system is fundamentally dependent on these variables. How could such a compromised system, which directly profits from the very problem that it's supposed to address, adequately move humanity in the right direction? This book points out that it isn't moving us in the right direction and that citizens will have to quickly reform our mass communication system to move us on the correct path.

The three areas that I focus on in this book are advertising, the digital infrastructure, and journalism, each uniquely move us in the wrong direction. Let's take a quick look at how and why each does this as a primer for the rest of the book. Advertising is easy. The simple purpose of commercial advertising, which is its most prominent form, is to encourage and increase consumption. Massive resources go into the research and development of advertising messages that try to persuade people to consume. The fact that this research exists suggests that elevated levels of consumption are not necessarily an inherent and intrinsic human behavior. If humans were to consume so many products and services naturally, then why would advertising be needed?

Nonetheless, constantly promoting products for people to purchase and consume is not necessarily a bad thing. However, if the production processes of the products and services to be advertised are not sustainable and they stretch the planet to its limits, causing harm, then this is a problem. This is where advertising gets complicated. Part of the challenge with advertising stems from its inception as a business transaction devoid of many social or environmental consequences. When organizations advertise, there are no regulations about the environmental impacts resulting from the transactions that enable the advertising. This means that the advertising system doesn't address the production processes and their relationship with nature for the products and services that are to be advertised. This is the heart of the problem with advertising and one of the main areas that I argue can be fixed.

Moreover, advertising is also deeply intertwined with the main mechanisms that propel the crisis forward. It is connected to production, consumption, and growth economics, all of which drive the global climate emergency. These mechanisms, which can be imagined as levers that trigger global warming, can be altered, and because they are so interconnected, altering one can change them all, which gives a reimagined and reformed advertising unique leverage in the battle for a sustainable future.

The infrastructure of the digital era, which is directly connected to advertising, also propels the world deeper into the climate crisis. The digital era's privatized infrastructure enables extensive commercial surveillance to a degree that has never been witnessed before. There is no precedent or historical comparison. Its goal is to increase levels of consumption that are already unsustainable as its infrastructure overexpends natural resources in its pursuit to convince us to overconsume. It uses the data it collects on us to individualize content and create highly informed persuasive messages to more effectively advertise to us. Moreover, the production and disposal of the devices that enable the internet, combined with the amount of energy required to power the digital era, make it unsustainable. Furthermore, the digital era contributes to inequalities by enriching a small but significant amount of millionaires who create elevated levels of carbon emissions. Many simultaneously use their wealth to lobby the government against measures that challenge the status quo by addressing the climate emergency. The idea that the digital era is safe because it's paperless and therefore doesn't harm the planet is wrong. As the digital chapters in this book demonstrate, multiple aspects of the digital world are moving the world in the wrong direction.

Besides advertising and the digital era, US commercial journalism is also challenging the progress that is needed to resolve the emergency. With very few news organizations and journalists left to write about any topic, US journalism is beyond a state of crisis at this stage. Commercial news media in the United States often ignore the causes of the climate crisis, downplay the solutions, disconnect weather patterns from global warming, or omit the topic all together. Coverage is often riddled with false information and misinformation, which delays the much-needed collective action required to address the crisis. Moreover, news organizations are predominantly funded by commercial companies, including those within the fossil fuel industry, which advertise and, in turn, promote the consumption of their goods and services. This means they are bankrolled by the same entities that have everything to gain from preventing the large-scale action needed to effectively address the climate crisis. We live in a consumer economy and news media are part of this world, extend it, and are often primarily funded by it.

With remaining news organizations barely staying afloat, it's not a good idea to run stories that damage relations with advertisers since they can easily pull their funding. Moreover, with the US news media system based on a libertarian philosophy free from government intervention, as opposed to a philosophy of social responsibility—where the press's responsibility is to inform citizens to help facilitate democratic governance—news organizations will be less likely to support government intervention. This means that calling for and supporting the governmental regulation required to effectively address the climate crisis would also open up the news industry to reform and more public oversight. Democratizing the news industry would mean creating more noncommercial, public, and local news outlets that would bite into commercial news organizations' market shares, which is a scenario they resist. In essence, the path that is needed to support mitigating the climate emergency clashes with many aspects of US commercial journalism.

Journalism, but also advertising and the digital infrastructure, exist within and reinforce growth economics. Growth economics guides businesses along with the overall economy and their goals. It does this by codifying a never-ending quest for profit at the expense of the planet by fueling constant overproduction and overconsumption. Infinite economic expansion on a finite space, which is our planet, harms the Earth by overstretching its ecosystems. Growth economics eventually fuels global warming and the climate crisis, which means it must be rethought or eliminated entirely as part of the battle for a sustainable world. Discussing this within the media is challenging because almost every aspect of US mass communication exists within and perpetuates this paradigm.

What do the sum of the challenges facing advertising, the digital infrastructure, and what's left of journalism mean for those interested in addressing the climate emergency? It means that *the fight against global warming is also a battle for media reform*. This means reforming advertising, the digital media infrastructure, and journalism. This book rethinks these three areas to help address and mitigate what many scholars and scientists consider the most important crisis of this century.[7] We need active democratic citizens to help reform our mass communication infrastructure to aid our transition into a sustainable world. There is an immediate need for new ways of thinking and understanding how mass communication contributes to this crisis, and how it can help humanity limit its worst effects. It appears there is no one fix to the global climate emergency, and multiple structures within societies will need to change in order to minimize the trajectory of the potential disaster.

Indeed, successful environmental movements will strive to dismantle or reform institutional infrastructures that facilitate the prolongation of the planetary global warming crisis and the unsustainable consumption of natural resources. Mass communication infrastructures need to be included in this goal. Reforms are necessary to achieve climate justice, but the urgency does not always make them politically salient. Changing powerful institutions to promote a sustainable future will not be easy. The climate crisis requires unprecedented and new forms of thinking, collaboration, and labor if it is to be addressed. Existing solutions aren't effective enough. New ways of organizing the world are needed immediately and that means rethinking the way institutions interact with nature. Transitioning out of the present political, economic, and mass communication systems calls for imagining, building, and realizing a new world through creative and inventive praxis. Humanity will need better media and advertising systems that contribute less to the climate crisis. It will also need to enable more democratic exchanges of information to successfully build a transition out of a system that may spell the end of the Anthropocene. Let's not underestimate this crisis—it is a deadly and serious emergency. The World Health Organization explains that the effects of climate change will directly cause approximately 250,000 human deaths per year starting from the year 2030.[8]

This book provides a potential path around some of these barriers and also outlines a plan to transition out of present mass communication policies that support the status quo. Its main focus is on the United States, since the United States is one of the worst historical contributors to the climate emergency. The United States is also the world's most powerful barrier to reform that can help address the crisis. However, many of the issues described in this book also pertain to countries all over the world, while many of the suggested reform measures can also be applied to systems outside of the United States. Therefore, while the content is somewhat local, it has immense global ramifications and many of its arguments are universally relevant.

Chapter Descriptions

This book is divided into four sections. The first section highlights the science behind the climate emergency and describes the political will (or lack thereof) to address it. The second section focuses on advertising, global warming, and reform. It contains three chapters, with each emphasizing a different aspect of the relationship between advertising and global warming. Because advertising appears to contribute most powerfully to the climate crisis, compared to other areas within the mass communication field, much of this book is devoted to discussing this influence and how it can be reformed. These chapters provide the empirical underpinning necessary to justify and develop the use of the advertising system as a positive means of mitigation to address the climate crisis. The third section focuses on the digital communication infrastructure, its technologies, and how they contribute to the climate emergency. It contains two chapters and each one discusses different aspects that contribute to the crisis. The fourth and last section of the book focuses on journalism and its relationship with the climate emergency. It contains three chapters, with each one noting significant challenges that delay action, while the last chapter addresses reform measures that would help address the crisis. The section below goes into further detail about the content of each chapter.

Chapter 1 outlines the science behind the climate emergency and discusses how greenhouse gas emissions are created, along with how they contribute to global warming. Citing multiple Intergovernmental Panel on Climate Change studies, there is an urgency to immediately and substantially lower GHG emissions in order to limit global warming to 1.5 degrees Celsius (C). Anything over a 2 degree increase will almost certainly set off massive damage by surpassing a tipping point beyond which the Earth's climate will continue heating, even if emissions were reduced. Scientists are also telling us that the window of opportunity to reverse the worst effects of global warming is rapidly vanishing and that governmental policies are the best way to address the crisis. Unfortunately, countries around the world, and especially the United States, are not doing enough to adequately address this crisis of epic proportions.

Chapter 2 discusses the US political will to address the climate crisis. It notes that wealthy industries, like the fossil fuel industry, have a long history of influencing the government to create legislation in their favor. It also discusses the influence of the neoliberal era, which promotes smaller governments and encourages laissez-faire regulatory policies while placing private, profit-oriented businesses as important power brokers. The administration of President Donald Trump magnified this approach, while ushering in a fossil-fuel coup d'état by allowing the industry to essentially directly run the governmental agencies that were designed the protect the public from the harm that those very industries create. In effect, the political will to address the climate crisis in the United States is very low, and concerned citizens will have a steep uphill climb in their efforts to reform and create sustainable climate policies.

Chapter 3 begins to focus on advertising, which exists as the mass communication field's most pronounced contributor to the climate crisis. It offers an underexamined history of advertising via its relationship with the climate emergency and discusses how its reform may be a useful tool in the battle against global warming. Advertising had a minimal impact on society during the precapitalist era, but then went on to grow in a mutually dependent and symbiotic relationship with each wave of industrialism and capitalist production capabilities. During this transition, advertising became subsumed under capitalism, which developed it into a pure business transaction devoid of any environmental or social implications. During industrialization, it became a powerful engine to propel forward a new consumer society that would help stretch the limits of the Earth's productive and regenerative capacity by encouraging the overconsumption of goods and services with environmentally harmful and unsustainable production processes. In this context, advertising became ideological, subservient to the capitalist mode of production. It reflected a dominance over nature, at the expense of environmental sustainability.

Chapter 4 continues the in-depth look into advertising and its relationship with the climate emergency. The goal of this chapter is to identify and understand how advertising interacts with the main mechanisms (i.e., production, consumption, and growth economics) of human-made climate change. Understanding how they work together helps us better explain the causes of the climate crisis, while laying a foundation for new policies to assist in reversing the crisis's trajectory. The chapter argues that the way advertising interacts with these mechanisms suggests that they are *interdependent* to such an extent that altering the function of any one of the four mechanisms would change the trajectory of the climate crisis. The second argument, which is premised on the first, maintains that advertising is a powerful influence on global warming and that, if modified, it would have a significant positive mitigative effect on the crisis. Modifying existing advertising systems can work as an additional strategy to address climate change. This means that activist, civic, environmental, and political groups interested in addressing the climate emergency may want to expand their mitigative strategies to include reform of the advertising system.

Chapter 5 rethinks advertising to present a mitigative model for environmental sustainability. It describes the beginning of a proposed model that will hopefully ignite a larger conversation about how reforming advertising can help in the battle for a sustainable future. A mitigative advertising model would use the advertising system as a gatekeeper through a system of fees to incentivize manufacturers to produce more environmentally sustainable products and services that cause less harm to the Earth's ecosystems. As such, a mitigative advertising model would promote products and services that are more likely to maintain a healthy and sustainable planetary ecosystem.

Chapter 6 moves on to the postindustrial advertiser-driven digital era and analyzes the tactics and technologies of commercial surveillance, with particular attention paid to online price customization. I argue that these strategies exploit online users and erode consumer agency, while exacerbating preexisting economic inequalities. The chapter opens with a description of the debate among scholars who focus on the internet's relationship to power, class, and economic inequalities. I detail the astonishing growth in online data collection, advertising, overall consumption, and e-commerce revenue during the past few decades. The private sector collects and sells user data to businesses and advertisers, who in turn use the data to individualize content in order to more effectively advertise, set prices, and sell goods and services. This surveillance function aids in the enhancement of consumption by acquiring data on consumer habits and preferences, which then allows advertisers to better target potential consumers with highly informed persuasive messages.

Chapter 7 explains how the advertiser-driven digital era undermines efforts to address the climate emergency. First, the entire premise of the digital era is to surveil online users to more effectively create individualized messages in order to increase consumption. Global consumption has tripled during the past 30 years. This momentum ignores the science that tells us consumption needs to decrease and not increase in order to address the climate emergency. Second, the production and disposal of electronic items used to support online platforms contribute to the overdraining of natural resources that helps fuel the climate emergency. Third, the amount of energy required to power the growing digital era is ultimately unsustainable. Fourth, the digital era creates and enriches a small but significant percentage of the global elite, which negatively impacts efforts to address global warming, as these groups contribute an extremely elevated percentage of carbon emissions through their consumption practices. Last, the macro effect of the digital infrastructure exacerbates economic inequalities while enhancing industry power and does little to enrich democratic governance, as those with more capital generally have a disproportionate political influence. This is problematic, as mitigation efforts have to go through governmental legislative processes in order to be successful, and the groups with the most power and influence are opposed to the structural changes needed to address the crisis.

Chapters 8, 9, and 10 move on to focus on the relationship between journalism and the climate crisis. Chapter 8 begins by discussing several US-based trends in commercial journalism that negatively affect climate change reporting and challenge democratic processes as they contribute to the delay of crucial political policy decisions urgently needed to address the climate crisis. It discusses how false balance, media silence, advertising interests, profit orientations, editor preferences, and other variables contribute to this challenge. These trends are concerning because the news media are the main source of scientific information for the public.

Chapter 9 continues to discuss influences that limit the quality and quantity of climate change coverage. It further examines the relationship between the fossil fuel industry's misinformation initiatives and the mainstream media. It describes how the myth of the liberal media, the commercialization of the media, the politicization of climate science, the cocooning of the internet, and additional challenges within political and economic systems that create massive inequities and disenfranchisement all contribute to the lack of public understanding and salience of climate change information and misinformation.

Chapter 10 begins by diving into literature that describes commercial media structures as a "cultural apparatus," which helps explain why the quantity and quality of climate science reporting is severely constrained in the United States. It is funded by companies that promote and benefit from overconsumption, which means stories that challenge those topics can damage relations with advertisers. The media are also owned by wealthy corporations and run by elites, who have transformed the corporate media into outlets conveying information that most often benefits other elites. This is challenging because elites consume the most and have the most to lose in a carbon-conscious world. Last, the US news media exist within a predominantly libertarian laissez-faire framework, which can clash with promoting the government regulation required to adequately address the climate emergency.

The challenges surrounding climate change journalism suggest this is a structural problem, especially since many news organization modes of operation often clash with realistic solutions to the climate crisis. Therefore, in order to begin to significantly improve climate change reporting, society will need to support media reform initiatives. There are significant differences between US and European news media structures, and the publicly funded European models are superior on climate change coverage, even though they are far from perfect. Democratizing the US media system will help diversify the information needed to inform the US public about significant issues that affect their lives. Access to relevant information is a founding tenet of a democratic society, and with the challenges facing the US media today, US democracy is in a deep crisis. Democratizing the US commercial media system to include more nonprofit, noncommercial, and public news media will begin to disconnect many of the challenges that journalists face when reporting about climate change. Calling for the multiple kinds of immediate government regulation needed to more adequately address the climate emergency would be more likely to occur under a reformed and democratized US media system.

I

Climate Emergency and Political Will

CHAPTER ONE

The Last Generation to Stop the Disaster

This chapter describes the science behind the climate emergency. Having a general understanding of the basic science behind climate change informs us about the severity of the crisis, which helps us as we move into how mass communication interacts with the climate emergency. The science tells us that climate change is already happening and that it is negatively affecting multiple properties of our planet. It also explains that there is a specific threshold of planetary warming, and any heating beyond that will be catastrophic. In order to decrease planetary warming, the world must reduce its GHG emissions, which will require significant reform measures. Moreover, the window of opportunity to do that is quickly diminishing. The end of this chapter briefly outlines how climate science interacts with the main mass communication themes that the rest of this book discusses in the additional sections. Unfortunately, science is quite clear that the world, and in particular the United States, is not doing enough to adequately address this crisis.

The ecological crisis,[1] perhaps now the most pressing problem facing humankind, is overwhelmingly attributed to human activity.[2] It has manifested itself through the warming of the planet, referred to as global warming. Its causes are deeply embedded in our political and economic systems through their conceptions and use of transportation, industry, heating, and other fossil fuels that support the production and consumption of products, services, and energy. Constant expansion through growth economics propels these forces forward. Together they create GHG emissions that increase the planet's temperature.

The most common GHG emissions are carbon dioxide, tropospheric ozone, methane, nitrous oxide, and chlorofluorocarbons, which are either exclusively linked with human industrial development or with an expanding human population that must be fed, which creates methane and nitrous oxide, and fuel, which creates carbon dioxide and tropospheric ozone.[3] Carbon dioxide (CO_2) accounts for 65 percent of global GHG emissions as a result of the burning of fossil fuels. Professor Justin Lewis of Cardiff University notes that these gases

> have the capacity to form a layer in the atmosphere around the Earth and prevent heat escaping. This, in turn, causes the planet's temperature to rise—a fairly straightforward process climate scientists call "the greenhouse effect." Evidence suggests that we have passed the planet's ability to absorb "greenhouse" gases at a sustainable level. Indeed, human activity has reduced the ability of the Earth to absorb gasses like carbon dioxide—by cutting down forests to make room for increases in the production of commodities—while relentlessly increasing the production of those gases.[4]

If all greenhouse gas emissions immediately stopped, climate change would still continue because it lags in the Earth-atmosphere system. If the system that creates the production of these gases goes unchecked, people and ecosystems risk irreversible damage in the coming decades.[5]

A 2018 Intergovernmental Panel on Climate Change (IPCC) report, along with hundreds of other reports throughout the past several decades, have been sending out warning signals to the world about the immediate need to mitigate global warming in order to limit the scope of the climate catastrophe.[6] The 2018 IPCC report urged immediate and unprecedented changes to global policy to keep global warming to a maximum of 1.5 degrees Celsius, given that a 2 degree Celsius increase will likely more negatively affect the poorer areas of the world compared to the "more developed" countries. This will be extremely challenging because estimated anthropogenic global warming is currently increasing at 0.2 degree Celsius per decade due to past and ongoing emissions. Global warming is likely to reach 1.5 degrees Celsius between 2030 and 2052 if it continues at its current rate.[7] The global average temperature has already reached 1 degree Celsius above preindustrial times and an additional warming of 0.4–0.5 degrees Celsius is expected as a consequence of the GHG emissions that have already been emitted.[8]

A 1 degree Celsius increase above preindustrial times is still viewed as problematic according to leading climate scientist James Hansen and a team of international experts.[9] A 2 degree increase will likely cause off-the-charts damage. Once the global temperature rises past the threshold of 2 degrees Celsius above preindustrial levels, self-reinforcing feedbacks will likely develop, which could drive further warming and prevent climate stabilization. This would mean continued warming toward a "Hothouse Earth" even as human emissions were reduced. The tipping point that initiates the "Hothouse" suggests it's a point of no return for humankind.[10] The climate would become out of human control with the creation of an unstoppable process.

Many global politicians are not doing enough to address this crisis. Sir Robert Watson, the United Nation's (UN) former top IPCC scientist, found that the 2015 UN climate agreement in Paris fell short of preventing a temperature rise of 1.5 degrees Celsius.[11] He noted there is literally no chance of making the 1.5 degree target under that agreement. The 1.5 degrees Celsius target has almost certainly already been missed due to inaction to stop the increase in global GHG emissions during the past 20 years. This is even more troubling as the world's population is expected to increase by 40 percent, or to 10 billion, by 2050, which means it will demand even more from the already exhausted planet.[12]

Kevin Anderson, chair of energy and climate change at the Tyndall Centre for Climate Change Research, agrees with the pressing challenge of this warming scenario unless radical changes are introduced. "I think it is quite reasonable to think we are heading to about 4 degrees centigrade of warming across this century. That would be utterly devastating. Just remember the difference between now and an Ice Age is about 5 degrees."[13]

He notes that without major changes, the world will experience more famines, droughts, floods, changes in food patterns, tensions within communities, and the breakdown of many ecosystems that pollinate crops and make air clean. The IPCC notes that climate-related risks to health, food security, water supply, human security, livelihoods, and economic growth are all projected to increase with global warming of 1.5 degrees Celsius and increase further with 2.0 degrees Celsius.[14] Disadvantaged, vulnerable, indigenous, and less developed countries will disproportionately experience higher risk of adverse consequences with global warming, while poverty and disadvantage are expected to increase as well. The ongoing crisis suggests that the Earth is no longer able to support the present levels of consumption and economic growth that have stretched the planet beyond its limits.[15]

Additional Effects of Climate Change

With the 1 degree Celsius increase, many land and ocean ecosystems around the world have already changed, to the extent that 82 percent of their core ecological processes on land and sea have been altered.[16] More extreme weather and climate-related events are expected to increase along with changes in average climate conditions, which are anticipated to continue damaging infrastructure, ecosystems, and social systems, according to the United States Global Change Research Program.[17] Besides increases in sea level and more droughts, climate change is also already affecting human food supply, health, productivity, and disease transmission.[18]

According to a recent study by the InterAcademy Partnership, the global food system is already broken as it leaves billions of people either underfed or overweight.[19] Simultaneously it is driving the planet toward climate catastrophe as it creates a third of all GHG emissions, which is more than all emissions from transport, lighting, heating, and air conditioning combined according to a study by 130 national academies of science and medicine across the world. Besides harming the food supply, climate change is also expected to cause growing losses to infrastructure and property, impede the rate of economic growth over this century, and increase economic inequality.[20]

The sum of these threats to humankind led naturalist Sir David Attenborough at the 2018 UN Climate Change Summit in Poland to note that the collapse of civilization and the natural world is not too far away. "Right now we are facing a manmade disaster of global scale, our greatest threat in thousands of years: climate change. . . . If we don't take action, the collapse of our civilizations and the extinction of much of the natural world is on the horizon."[21]

According to Bill Gallegos, a Chicano activist and former executive director of Communities for a Better Environment, the ecological crisis is only one consequence of a system that's caused significant damage in the United States, and in many other areas throughout history.

> The ecological crisis began with the genocide of native peoples and the theft of their lands. That's when we started to burn down forests and do monocrop agriculture. The slave trade and slavery created the horribly destructive plantation system in the South. These plantations destroyed forests, ruined land, and despoiled the waters. The military annexation of the Southwest and California from Mexico

was the foundation for the ecologically devastating mining—coal, copper, uranium, silver—and petroleum-based economy. It also helped to stimulate the development of industrial agriculture systems, systems which super-exploited the labor of African American, Chicano and Mexican, Pilipino farm workers, and Asian and Puerto Rican farm workers in the Hawai'ian plantations.

There's a deep integration of the ecological crisis in this country with this history of racism and national oppression, a history whose legacy lives on today in the lives of Black, Brown, Native, and Asian communities throughout the United States. And I think what that means is that we can't solve these as separate questions. They are intertwined and integrated and they need to be solved together. So our program for ecosocialism or whatever we would call it must have at its center, for example, the genuine recognition of the sovereignty and treaty rights of the native peoples.

We not only need to stop oil drilling and fracking and build out a lot of solar energy, we need to restore and honor the land rights of the native peoples who over thousands of years have learned a very different relationship to mother earth. That's one path towards restoring the land and restoring our connection with mother earth, but every social movement of people of color and also poor white folks has an environmental justice component to it. The quality of the water that they drink, the soil that they live on, the air that they breathe, is affected by the fact that they're poor, or they are poor and dark skinned.[22]

Need for Rapid and Radical Change

Scientists agree that the window of opportunity to reverse the worst effects of global warming is rapidly vanishing and if governments do not act now, the window of opportunity will soon be gone.[23] If the human race wants to continue existing on the planet without the threat of its extinction, there is an immediate need for rapid and radical change in the way the world's systems work. This is the race and goal that we must all work for. Scholars John Bellamy Foster and Brett Clark put it more succinctly, "never before in human history has civilization faced so daunting a challenge."[24] Time is also running out. The next decade is crucial in minimizing large-scale and potentially catastrophic climate change.

While mitigation and adaptation efforts have substantially increased in the past few years, they do not yet approach the scale considered necessary to avoid substantial damage to the economy, environment, and human health over the coming decades.[25] Mitigation and adaptation can effectively address the worst of the climate crisis if governments are able to support them. Mitigation is the reduction of GHG emissions that are released into the atmosphere, while adaptation is the adjustment in human or natural systems in response to expected or actual climatic stimuli or their affects in order to moderate harm or exploit beneficial activities.[26]

Mitigation efforts are lacking. Eight months after the 1.5 degree Celsius target was set, it was already close to being broken, which indicates the world may come close to missing a key climate target. The Earth's average temperature is on track to increase by 2 degrees Celsius around 2050, according to Sir Robert Watson.[27] Drew Shindell, Duke University climate scientist and coauthor of a recent IPCC report, noted that "it's extraordinarily challenging to get to the 1.5C target and we are nowhere near on track to doing that."[28] James Hansen, the foremost US climate scientist, agrees the world is failing miserably at addressing climate change.[29]

The 2018 United Nations Environment Programme Emissions Gap Report suggests governments will need to urgently introduce new measures to stave off the worst effects of climate change.[30] The IPCC furthers this recommendation by noting governments around the world will need to take rapid, far-reaching, and unprecedented changes in all aspects of society if they wish to avoid disastrous levels of global warming.[31] This is because far-reaching, multilevel, and cross-sectoral climate mitigation would reduce future climate-related risks.[32]

In order to stay below 2 degrees Celsius, global GHG emissions should be reduced by 22 percent by 2030 from current levels according to the IPCC and the UNEP Emissions Gap assessment.[33] Science tells us carbon emissions need to be reduced between 50 percent and 85 percent by 2050. Overall emissions must be zero or negative by 2100. In essence, a tripling of efforts will likely be needed to keep warming to less than 2 degrees Celsius and only a rapid turnaround can help. In order to stay within the 1.5 degrees Celsius scenario, the world will need a fivefold increase in effort. If nothing changes, the world will see global warming of between 3 and 4.8 degrees Celsius by 2100, with warming continuing afterward.[34]

Moving Forward

Scientists propose numerous strategies to address the climate crisis. Countries will need to invest in clean technology, stop directly and indirectly subsidizing fossil fuels, create new taxes on fossil fuels, create environmentally sustainable production processes, decrease consumption, as well as initiate other strong government policies to bring down emissions. Planet-wide carbon-emissions reductions of a minimum of 3 percent a year is needed, with three times that number within rich, high per-capita carbon emitting nations.[35] Countries around the world may want to build publicly owned, decentralized, and democratically controlled power grids instead of energy systems that large corporations economically benefit from, especially those that burn fossil fuels.[36] The IPCC argues for international agreements to lower GHG emissions, while Oil Change International proposes keeping 80 percent of known fossil fuel reserves in the ground to sustain planetary life.[37] In addition, scholars Mark Jacobson and Mark Delucchi note that water, wind, and solar technologies can provide 100 percent of the world's energy.[38] However, countries like the United States would have a long way to go on this, as renewable energy constitutes only 9 percent of all energy sources, which is almost tied with nuclear electric power (at 8%). These two energy sources are at the bottom of the US energy totem pole.[39]

Climate scientist Kevin Anderson argues the only way to get emissions down is by reducing the current level of energy people use because investing in renewable energy sources and nuclear energy,[40] which would take close to 20 to 30 years, will be too late.[41] Immediate degrowth strategies with revolutionary change to the political and economic systems in the EU, the United States, and other wealthy nations may best address the crisis according to the Tyndall Centre for Climate Change Research.[42] The idea that the world will grow out of the crisis is false and misleading, as growth economics places the planet on a direct collision course with disaster by overstretching the planet's resources. The concept of degrowth, which has multiple definitions and interpretations, generally rejects the illusion of growth and advocates for democratically shrinking production and consumption with the aim of achieving social justice and ecological sustainability.[43] The concept of degrowth sees perpetual economic growth as the underlying driver of the climate crisis.

Need for Improvement

Countries around the world need to do a much better job addressing the climate crisis according to the 2019 Climate Change Performance Index. It ranks countries' climate protection based on 14 indicators within GHG emissions, renewable energy, energy use, and climate policy. Out of a total of 56 countries, no country ranked in the very high category. Here is the list of the top countries that did rank in the "high" category starting with the best: Sweden, Morocco, Lithuania, Latvia, the United Kingdom, Switzerland, Malta, India, Norway, Finland, Croatia, Denmark, the European Union, and Portugal. Coming in last were Ireland, Japan, Turkey, Malaysia, the Russian Federation, Kazakhstan, Canada, Australia, Chinese Taipei, Korea, the Islamic Republic of Iran, the United States, and Saudi Arabia. To compare the top two scores with the top bottom scores, Sweden achieved a 76.28 score with Morocco earning 70.48. The United States (second to last) came in with a score of 18.82, while Saudi Arabia earned a miserable 8.82. With the lack of political effort to address the crisis, some scientists are beginning to believe the world may have already crossed a series of climate tipping points, but they also acknowledge the science of tipping points is complex and that uncertainty remains.[44]

Importance of Mass Communication in This Battle

While climate science has succeeded in determining many causes of, and solutions to, the climate crisis, it's now up to a communication battle to convince the public about the immediate need to reform the way our societies are organized in order to address the crisis. The challenge with this scenario is that multiple aspects of our mass communication system are deeply intertwined with this crisis. The mass communication system consists of many of the same aspects of society's social organization that climate science says need to be changed in the first place. This hampers efforts to effectively address the emergency. Much of our mass communication system is owned, funded by, or profits from the very parts of the social organization that have to be changed to create a sustainable world. After chapter 3, the rest of this book delves more deeply into the multiple roles that various aspects of mass communication embody as they help move the climate crisis forward.

The three areas of mass communication that this book focuses on—journalism, advertising, and digital media—each interact with climate science in unique ways. Journalism and the mass media, along with social media platforms, influence public perceptions about science, as they generally constitute the main sources of public climate information. The way journalism organizations frame climate information and set the amount of coverage influences what citizens are supposed to think about and affects levels of public concern, which go on to influence how science is translated into public policy. This means that mass communication is at the heart of this process and that it is an extremely important player in the overall struggle for a sustainable future. Media, advertising, and the digital world are structures with values, power, preferences, special interests, and goals, which influence the way they discuss and interact with global warming.

The advertising system promotes endless consumption, allows mis- and false information to be distributed, fuels growth economics, funds news media, and helps enable climate countermovement groups funded by the fossil fuel industry. These aspects all influence how climate science is interpreted and understood. At the same time, advertising can also promote less consumption and sustainable consumption, expose false information, and help disclose important climate information to the public. Unfortunately, we are less likely to encounter these topics in our society.

While it contains both advertising and journalism, the digital world is also unique. It interacts with global warming science differently because it is mostly regulated in favor of the large companies that enable it.[45] This means it exists as a vast commercial surveillance system designed to extract data on human preferences, which are then used in individualized efforts to modify their behavior, most favorably with increased consumption that benefits industry. This means the organization and broad goal of the privatized digital world counters what climate science is telling the world, which is to *decrease* consumption. Moreover, the amount and kind of energy that powers the digital world is unsustainable.

The digital world is also a hotbed of false and misleading information critical of climate science. For example, social media bots spout propaganda by wealthy and powerful groups vying to manipulate public opinion by repeating content questioning science to confuse, misinform, and delay government action on global warming. Massive amounts of financial resources from the fossil fuel industry help create online content, while climate denial bloggers continue to receive best blog awards. Simultaneously, climate activist groups are shedding light on false information and propaganda by exposing the secretive fossil-fuel-based funding sources of blogs and other media. Nonetheless, the overall privatized structure, goals, lack of regulation, and unsustainable power supply make the digital world clash with the mitigative objectives required by climate science to address the climate emergency. This means that the battle against global warming is a fight for media reform.

As of 2019, many viewed the world as having roughly 10 years to immediately address the crisis, which placed this generation as the last that can prevent catastrophic global warming. It will also be the first to suffer some of the worst of its impacts.[46] We've reached a junction, which needs to be a transition, as this struggle over the future will determine how the planet will function. What we are experiencing is a pivotal moment in the history of the Earth. Will we choose extensive environmental harm or create something new and sustainable more akin to overall human development, with happiness, access to education, healthcare, and nutrition? Or will humanity continue with growth economics, individualism, and infinite consumerism all pushed and marinated through nonstop advertising, media, and marketing? Many of the answers to these questions are connected to political will power and the manner in which political systems function—both topics that are deeply connected to mass communication. The next chapter continues the discussion of the climate emergency within the US political system by examining recent trends under the administration of President Donald Trump. This fossil fuel–sponsored administration intentionally impeded the government regulation required to address the climate emergency in addition to other problems that harm human lives and the environment.

CHAPTER TWO

US Political Will to Address the Climate Crisis

This chapter is less about mass communication and more about the US political will, or lack thereof, to address the climate emergency. Citizens need to know who is in office, who funds them, and what they think. This information can help social movement strategies and also shows what the climate and media reform struggles are up against. I highlight how the fossil fuel and other wealthy industries have a long history of influencing governmental legislation in their favor. Much of this has increased during the past 40–50 years, during a period known as the neoliberal era. The transition into the Trump administration enabled the fossil fuel industry to take control over several governmental agencies. Their efforts resulted in an all-out attack on environmental regulation. Besides highlighting the power and hold that the fossil fuel industry have over the US political system, this chapter also describes some of the disastrous effects on human health. Toward its end, I connect mass communication with this chapter's theme to foreshadow what's to come in the rest of the book.

There is a long history of powerful groups resisting government regulation to address the climate emergency. Beginning in the late 1970s and early '80s, the wealthy and powerful began to see a pronounced economic and political threat to their power through industry regulation in response to climate change. This was happening even when every major US and multinational oil and gas company knew of the negative impact their businesses were having on climate change, according to the Pulitzer Prize–winning InsideClimate News.[1] They knew eventually the public would rise up against the harm their industries were causing the planet. However, the powerful and wealthy have a very long track record of distaste for governance where citizens are engaged and active, especially the kind that regulates markets and industries in favor of the public good. Working for the public good can mean creating regulations that limit cancerous industrial emissions, preventing industries from putting lead into toys, and breaking up oligopolistic markets so that consumers don't overpay for products. It can also mean regulating industry so that air and water are not polluted, while dismantling the fossil fuel industry to address global warming. Regulating the media so they work in the public interest also fits this approach. However, all of these regulations can cut into corporate profits and, as a result, many wealthy and powerful entities put forth efforts to challenge or eliminate them.

It is important to remember that many of the wealthy received their fortunes from within the fossil fuel industry, and their influence within politics and policies goes back well over a century.[2] In 2013, fossil fuel interests (gas, coal, and oil industries) in the United States spent almost $157 million in lobbying, which added up to over $430,000 per day, or $24,000 per hour according to OXFAM.[3] These lobbying efforts have paid off with substantial subsidies and tax breaks, which for the countries in the Organization for Economic Cooperation and Development outweigh those for the renewables sector. According to the International Monetary Fund, fossil fuel companies enjoy an estimated $5.3 trillion subsidy per year, the equivalent to $10 million a minute every day. This sum, which represents 6.5 percent of global Gross Domestic Product, is greater than the entire health care spending of all of the world's governments.[4] Much of this money—$2.2 trillion—could also be wasted by fossil fuel companies as they continue unnecessary exploration projects.[5]

Even while oil and gas corporations spend enormous sums to misinform the public that global warming does not exist, or at least label it as not human made, they don't have a problem asking taxpayers for additional money to protect their facilities around coastal areas from the effects of climate change. In one case that meant asking for $12 billion as an additional subsidy to the fossil fuel industry.[6] And when the oil industry asks, it usually gets what it wants fairly quickly. The US government fast-tracked an initial $3.9 billion to start building storm barrier projects that would specifically protect oil facilities from extreme weather and sea level rise.

Ronald Reagan's and Margaret Thatcher's Responses to Global Warming

The Reagan/Thatcher era is essential to highlight because it was a turning point that emphasized deregulation of the market in favor of capital, along with the privatization of industries and public goods. While President Jimmy Carter appeared to have initiated initial efforts to align large corporations with Wall Street's market metrics through a deregulatory program,[7] Reagan and Thatcher took this approach and multiplied it significantly. Their version would go on to attack the rights of workers, since they could get in the way of business profits and ownership demands. This transitional era became known as the neoliberal era and its approach promoted less government, which placed private, profit-oriented businesses as important power brokers. Businesses wanted to privatize public services to boost profits now that neoliberalization would open access to the money that previously went to government services. In essence, the supremacy of the market becomes paramount and not to be questioned during this era. Nevertheless, neoliberalism didn't just appear overnight; it is part of advanced capitalism where capital becomes less competitive and more concentrated in fewer hands. Ultimately, the neoliberal era reallocated power away from the public and toward capital, which also eroded democratic participation.[8]

The beginning of the Reagan/Thatcher neoliberal era influenced some of the initial governmental responses to global warming. Thatcher molded much of the early discussion of climate change during her tenure by diffusing responsibility away from Western countries. The idea was to make the problem a global challenge, for the entire world, as a means of deflecting primary and historic culpability from mostly Western countries and industries. The strategy downplays the reality that Western countries are the main contributors to historic GHG emissions that increase global warming and that many in these countries overconsume and live in luxury compared to most people around the world. But politically, in order to keep overconsumption and economic growth policies in place, it was convenient to create a situation where poor countries around the world would have to address the worst aspects of global warming in order to lower overall global emissions. Thatcher outsourced the responsibility onto Brazil and other third world countries (many of them previous English colonies) through identifying deforestation and other "far away" challenges.[9] It was important to make sure the ideological pillars of neoliberalism and growth economics would not be questioned. In fact, according to Thatcher, economic growth was needed in order to generate the wealth required to pay for environmental protection.[10]

The people driving neoliberal polices also tried to reframe how citizens were supposed to view themselves, their actions, public participation in governmental affairs, and the government. The idea of citizens participating and creating new policies, which are benchmarks of democratic processes, were downplayed during the Reagan/Thatcher era. Instead, the reframing of active citizens as passive consumers became one of the central themes. The US and UK media, often driven by elite sources, then began to shift toward portraying citizenship and citizens in more consumerist terms.[11] This is particularly troubling in low-diversity political party states like the United States because fewer alternatives are able to present themselves.[12]

Instead of the citizenry, private enterprises and private markets were to be considered the most important drivers of society, and therefore should be off-limits to anyone who cares about progress and the economy. As a result, citizens were supposed to engage in marketplace activities first and foremost. Engaging in politics or imagining oneself as an active citizen needed for democracies to function was not encouraged and was often relegated to the limited times every few years when people were able to vote. Participatory citizens are a direct challenge to the power that wealthy groups have over the political and economic system. From their perspective, it is best to convince the public that government is bad and not worth the effort to participate in. This approach also justifies government climate inaction, and it continued in the post-Reagan/Thatcher era, where the goals of both Bush administrations were to cast doubt on the science and exaggerate the costs of action to address climate change.[13] As bad as these administrations were to the environment, the Trump administration would top everything that came before it, creating the most dangerous, anti-climate-science administration in history.

Fossil Fuel Coup D'état in the Trump Era

It is not surprising that even before the Trump administration set foot in the White House, big business was already celebrating. An article on CNN online noted how "American CEOs are thrilled with Trump."[14] They predicted slashed tax rates and fewer regulations under the new administration, which would further enrich the top levels of many corporations. Profits would likely increase, as Trump, and many others in his cabinet, already had investments in multiple fossil fuel corporations. Trump and his businesses have multiple stakes in various energy-related projects with over $1 million dollars invested in the industry.[15] With Trump in office, the fossil fuel industry quickly became one of the greatest beneficiaries of the new administration.

However, working closely with fossil fuel industries is nothing new. US political administrations have a long history of attentively working with them. Connections between political parties, federal agencies, and the fossil fuel industry go back decades. For example, researchers have found that oil companies and the Republican administration of President Richard Nixon had manipulated information to diffuse opposition to offshore drilling and delegitimize protest groups in the late 1960s. Oil representatives also worked with the US Mineral Management Service to employ diversionary reframing tactics to change public debates on offshore drilling.[16] These connections flow between both Republican and Democratic administrations. Therefore, while it was egregious, it wasn't that surprising that immediately after Trump's inauguration he would go on to place top lobbyists and others connected to the fossil fuel industry directly in charge of environmental agencies.

Nonetheless, the degree to which Trump appointed industry lobbyists and purposely created climate misinformation was extreme. Trump is a climate denier, who has suggested climate change is a Chinese hoax and withdrew from the Paris climate agreement. He wasn't the only person to call it a hoax, as Republican Senator James Inhofe, who at one point chaired the Environmental and Public Works Committee, called climate change "the greatest hoax every perpetrated on the American people" back in 2003.[17] However, adding China into this falsehood was a new and convenient way to rile up his base to garner support for his trade war against the country. Accusing China for US economic problems would be a great way to deflect blame for an extremely unequal economic system that primarily benefits the wealthy at the expense of everybody else. Many politicians who suggest climate change doesn't exist may not actually believe in what they are saying, but they do realize that supporting that position has political capital. According to James Hansen, "Many of the conservatives know climate change is not a hoax. But those running for president are hamstrung by the fact they think they can't get the nomination if they say this is an issue. They wouldn't get money from the fossil fuel industry."[18]

The fossil fuel industry's political investment paid off in a big way under Trump. He appointed climate denier Myron Ebell,[19] director of the Competitive Enterprise Institute, a conservative libertarian think tank, which is partly financed by the coal industry,[20] to oversee the transition plans for the Environmental Protection Agency (EPA). Trump then picked Republican climate denier Scott Pruitt, attorney general of Oklahoma, to lead the EPA. He was part of a major legal action against the EPA to block efforts by the Obama administration to curb GHG emissions from coal-fired plants.[21]

Andrew Wheeler, an ex-coal lobbyist, would become the acting administrator at the EPA after Pruitt's ousting for corruption. Then in early 2019, Trump selected David Bernhardt, a former mining, oil, and gas lobbyist, to run the Interior Department. He was expected to continue the Trump administration's priorities to further advance oil and gas interests, especially on or near public land.[22] The Senate confirmed him for the position in April 2019.

In a 52–45 vote, the Senate also confirmed Jeffrey Bossert Clark, who defended oil giant BP (formerly British Petroleum) from lawsuits over the 2010 Deepwater Horizon oil spill, as the administration's top environmental lawyer. Democrats Claire McCaskill and Joe Manchin voted with Republicans to confirm him as assistant attorney general for the Environmental and Natural Resources Division.

Clark has repeatedly challenged the science of climate change, calling it "contestable."[23] The initial purpose of his position was to bring cases against companies and individuals when they break civil and criminal anti-pollution statues. Trump's Presidential Committee on Climate Security also included climate "doubter" and Princeton physicist William Happer, who has suggested higher levels of carbon dioxide emissions are beneficial.[24] Happer's role was to create some form of "legitimacy" for the administration by doubting and critiquing climate science produced by areas within the government that could contradict the Trump administration's stance on denying the existence of human-made global warming. This method would help to slow, censor, and put pressure on government employees to fall in line and follow Trump's war on climate science.

Trump's first secretary of state, Rex Tillerson, was also connected to the oil industry as a chief executive of ExxonMobil, while his national security advisor, John Bolton, openly spoke of taking over Venezuela's oil industry. He said "it will make a big difference to the United States economically if we could have American oil companies really invest in and produce the oil capabilities in Venezuela. It'd be good for the people of Venezuela. It'd be good for the people of the United States."[25] According to Andrew McCabe, former acting FBI director, Trump also asked why the United States wasn't at war with Venezuela, noting "they have all that oil and they're right on our back door."[26] Trump went on to place at least seven more climate deniers to guide every governmental agency that dealt with climate change. Not surprisingly, with the top environmental agencies in the US government headed by fossil fuel industry lobbyists, the Trump administration was failing to enforce the environmental rules that remained on the books, according to a watchdog group.[27] At the same time, the Trump administration also focused on eliminating offshore drilling regulations that were put in place to prevent another massive catastrophe like the lethal BP Deepwater Horizon disaster in the Gulf of Mexico by Louisiana.[28] It soon became clear there was a fossil fuel industry takeover of the institution that was supposed to protect the people and environment from the very industry that was now running the institution.

In effect, climate denial would become the official White House doctrine, and it appeared as if a fossil fuel coup d'état had commenced. The industry and their climate denial groups had someone to give them direct control over policy. Trump's senior EPA officials, once in office, would go on to work very closely with the Heartland Institute, a conservative think tank that rejects the science on global warming and dismisses climate change.[29] The American Petroleum Institute, which is the leading lobby group for US oil and gas companies, also sent the EPA a written top priority wish list called "comments on specific regulations." Trump's people worked so diligently on the list that by the end of Trump's first year in office, his EPA had already fulfilled almost all of the American Petroleum Institute's major deregulatory demands.[30] Scott Pruitt's released schedules showed a majority of his meetings were with fossil fuel industry stakeholders, while only 1 percent of his meetings were with environmentalist or science groups.[31] He rolled back standards for power plants, cars, pesticide restrictions, wildlife protections, and even school lunches for children. Trump's goal was to gut a government agency responsible for protecting nature, humans, and animals from pollution and other harmful elements to boost corporate profits.

Government-Sponsored Censorship

It was clear these radical, fossil-fuel industry people were dismantling as many pro-environmental business regulations as possible while cutting crucial public climate change research funding. They wanted the government to suppress scientific information and explanations about climate change from the public. However, in order to do that, the government needed to censor information and voices within several governmental agencies. It began by eliminating scientists, groups, and programs within government that would challenge the Trump administration's false narrative on climate change.

Once again, this was nothing new. Even before Trump, Republicans and think tanks used various methods of intimidation to control scientists in order to resist the science of climate change. For example, a Republican-led committee in the House was accused by climate scientists of politically motivated "harassment" during their investigation into the National Oceanic and Atmospheric Administration.[32] In addition, a conservative climate denial think tank, the Competitive Enterprise Institute, had also sued climate scientists.[33] After the father of climate science, James Hansen, told Congress in 1988 that reductions in the burning of coal, oil, and other fossil fuels were necessary, his employer—the National Aeronautics and Space Administration or NASA—suddenly appointed a media overseer who vetted what Hansen said to the press. Hansen also claimed the White House altered his subsequent testimony.[34]

Trump's approach was to first target government employees who were involved in climate agreements. The environmental group Friends of the Earth acquired a questionnaire Trump's transition team circulated at the Department of Energy requesting the names of employees who had attended climate talks over the past five years.[35] Pruitt then decided not to renew half of the board of scientific counselors, which advises the EPA on quality and accuracy. He replaced them with representatives of the industries the EPA was supposed to regulate.[36] The committee carrying out the Sustained National Climate Assessment was also eliminated by the head of the National Oceanic and Atmospheric Administration at the beginning of the Trump administration.[37] Part of the advisory committee's mission, which was to help state and local governments, federal agencies, and the private sector understand the science and data behind climate change, appeared too much of a threat to the Trump administration.

Trump's EPA also started censoring government workers by canceling the appearances of at least three of its scientists who were set to speak at a Rhode Island conference in October 2017.[38] The White House also tried to censor a State Department senior intelligence analyst from discussing climate science in congressional testimony because the analyst did not match the Trump administration's views.[39] Trump's National Climate Assessment was also slated to work through a multiagency effort to undermine climate science by removing additional content in upcoming reports. In addition, the National Security Council under John Bolton told employees to remove references to climate change in speeches and public statements, while the US Geological Survey would no longer project the impact of climate change past 2040.[40] According to *Science* magazine, the administration also eliminated NASA's Carbon Monitoring System, which made it more difficult for the government to measure the levels of carbon dioxide and methane in the atmosphere.[41] Under these circumstances, government workers' jobs are at risk if they speak out, which creates an atmosphere not only of direct censorship but also of self-censorship.

Another way the Trump administration tried to control climate narratives was to release climate information on holidays, place misinformation in agency reports, take charge of the dot.gov website, and implement additional bans through numerous governmental agencies that either interacted with, or discussed, climate change. The Trump administration would release an important study from the US government regarding the human impact on climate change during Black Friday, where Americans are encouraged to shop and consume. This is not a new communication strategy; every administration does this. Anyone who works in public relations knows that the best way to bury a story is to release it during a holiday, crisis, or other event like Black Friday, where it'll likely receive little media attention. The irony behind Black Friday is thick given that a consumer society is one of the main causes of global warming. When asked for a reason why they decided to release the study during this shopping holiday, the administration declined to answer.

Trump also placed Indur Goklany as a senior advisor at the US Interior Department's Office of Policy Analysis. Goklany would go on to place misinformation and multiple false claims about the climate crisis into several agency reports. Not surprising, Goklany has multiple connections to conservative climate denial think tanks like the American Enterprise Institute and the Cato Institute. The *New York Times* found that Goklany's revisions suggested that carbon dioxide is beneficial for the Earth's atmosphere and that there still wasn't a scientific consensus that the Earth is heating.[42]

The administration also began to systematically cut references to climate change on government websites—sometimes the censorship was successful, other times it appeared as if only a few mentions were deleted. For example, the National Institutes of Health, which is considered the world's leading public health research body, found multiple references to climate change deleted on its websites.[43] Employees at the US Department of Agriculture were also told to not use the term "climate change" in their work, and instead use "weather extremes." However, that was just the start of the end of several other terms. The term "reduce greenhouse gases" was also censored in favor of "build soil organic matter, increase nutrient use efficiency." Other terms and references about economic growth and emerging business opportunities were to be tolerated and appreciated by all on the staff.[44] Another Trump administration tactic was to simply rename fossil fuels as "molecules of freedom," in particular natural gas produced in the United States.[45] The idea here is to rebrand natural gas, which is predominantly made of methane, with magical abilities that will spread freedom around the world for those who purchase it.

These forms of censorship also happened at the state level under Republican and conservative Democratic governors. Under Wisconsin Republican governor Scott Walker, the Department of Natural Resources removed language from its public website that described the scientific consensus that human activity is the main cause of climate change.[46] In Florida, which is the state most susceptible to the effects of global warming, Republican governor Rick Scott's Florida Department of Environmental Protection was ordered not to use the terms "global warming" and "climate change" in any official communications, emails, or reports.[47] Under conservative Democratic governor Bev Perdue, North Carolina passed a law forbidding planners from taking future sea level projections into their planning decisions, even though they are a coastal state.[48]

Besides censoring climate information from the public, many Republicans decided they needed to go ever further to support the fossil fuel industry. It wasn't enough to regulate in favor of the oil industry, they also felt the need to eliminate resistance to its objectives. The Trump administration, filled with ex-fossil fuel lobbyists, would be the perfect group to criminalize this resistance. The US Department of Homeland Security would go on to describe a group of Climate Direct Action supporters as "suspected environmental rights extremists" and listed two of their members alongside violent white supremacists who engaged in mass killings.[49] These supporters, known as the "Valve Turners," engaged in nonviolent climate civil disobedience by temporarily closing a few oil pipeline valves while waiting patiently to be arrested. They also informed the energy company before their direct action.[50]

To preempt any such disruptions, the Trump administration proposed legislation that would prescribe up to 20 years in prison for conspiring to inhibit, or for inhibiting, the operation of an oil or gas pipeline.[51] At least 17 states have introduced similar bills since Trump was inaugurated. This legislation would harm efforts in the climate movement for simply providing training or logistical support for peaceful protesters. Meanwhile, across the sea, the United Kingdom counterterror list, which is used throughout England as an antiradicalization scheme designed to catch those at risk of committing terrorist violence, also included nonviolent environmental groups such as Greenpeace and Extinction Rebellion alongside neo-Nazi and jihadist groups.[52]

Gift to Industry/Harm to the Public

Trump's gift to fossil fuel lobbyists came at the expense of the health and longevity of not only the environment but also of human lives. His environmental policies were extremely dangerous, as they endangered human and nonhuman lives alike, as well as brought harm to public health-care systems. According to an article in the *Guardian* newspaper (UK), "the mission of the Pruitt EPA seems to be maximizing polluting industry profits *at the expense* of human health and the environment."[53] The EPA would go on to leave no stone unturned; even drinking water would be deregulated. By early 2019, Trump's EPA was expected to refrain from setting limits on harmful chemicals found in drinking water, even though many of those chemicals are the same found in Teflon and Scotchgard, which scientists have linked to a number of diseases including cancer.[54] Trump's EPA also rejected a proposed ban on a widely used toxic pesticide linked to brain cancer in children. By July 2019, it was preparing to deregulate nuclear plants by cutting the size and scope of radiation-protection and emergency preparedness inspections at the plants.[55] Moreover, his administration also worked to eliminate regulations that required the fossil fuel industry to detect and plug methane leaks at oil and gas facilities. Methane is extremely destructive to the atmosphere, and the administration's decision occurred after the release of a study by the Environmental Defense Fund that concluded that the oil and gas industry was underreporting the extent of its methane emissions to the EPA.[56]

A recent study by two Harvard social scientists suggested the Trump administration's rollback of environmental regulations could cause the deaths of an additional 80,000 or more people per decade and cause an additional 1 million Americans to contract respiratory illnesses.[57] According to the Environmental Integrity Project, multiple US oil refineries are emitting levels of cancer-causing benzene that are above the federal government's limits. The number of harmful chemicals that are emitted from these facilities disproportionately harms low-income people and minorities, often by worsening their asthma and by influencing other respiratory problems.[58] Almost half of the US population is already breathing unhealthy, heavily polluted air according to a study by the American Lung

Association,[59] while the World Health Organization notes some seven million people die each year from air pollution alone.[60]

The economic crisis caused by the Covid-19 pandemic was another great opportunity for Trump to bolster the fossil fuel industry and other large corporate polluters while further risking American lives. After the American Petroleum Institute sent the EPA a letter calling for regulatory suspension, Trump used the pandemic to suspend environmental rules, effectively telling companies they will be allowed to regulate themselves during the crisis. This meant that large polluters would no longer be held accountable or punished for failing to comply with government regulations. Companies could now lawfully pollute the air and water more than before. As a bonus, they wouldn't be required to monitor their actions, effectively undercutting efforts to know how bad the polluting may have been.[61] To make things even worse, Trump then weakened rules to compel auto companies to produce more fuel-efficient vehicles. These changes would allow vehicles to emit about a billion more tons of heat-trapping carbon dioxide, while forcing Americans to spend more on gasoline. It will also lead to more life-threatening air pollution.[62]

Pollution weakens lung functions and can also play a role in transmitting viruses. The risk of death is often much lower from viruses in less polluted areas. This means Trump's decision would likely increase the already high American death toll from Covid-19 given that studies indicate that even small increases in air pollution can lead to an increase in Covid-19 death rates.[63] What is the Trump administration's response to this crisis in air quality? Besides eliminating regulations designed to protect people, his plan was to change how the government counts the number of people that are killed by air pollution, which would dramatically undercount premature deaths due to dirty air.[64]

These are crimes against humanity. Those in the climate justice movement were right to be in shock with the thought of a Trump presidency, while the world's scientists were also correct to declare the Trump presidency a "disaster for the planet."[65] This sort of impending and unseen violence needs to be resisted. One has to ask, will any Republicans in Trump's party stand up to his administration's assault on the planet?

US Partisan Politics

The Republican Party is probably the last remaining political group in the developed world that denies climate change is human made. Famed

US linguist and dissident Noam Chomsky recently called the Republican Party the most dangerous party on Earth, and criminally insane for their climate denialism regarding the threat global warming poses to humankind.[66] Some Republican politicians are so radical and devoted to corporate profits over government regulation to address the climate crisis that they would prefer to refuse to do their jobs and abandon the democratic process to prevent any sort of progress that would address the emergency. Oregon Republican senators did this by fleeing to Idaho in response to incremental, market-based state legislation introduced by the Democrats in 2019. Responding to an order for the police to bring them home, Republican senator Brian Boquist told reporters "send bachelors and come heavily armed. I'm not going to be a political prisoner in the state of Oregon."[67] His call for potential violence over regulation resonated with armed right-wing militia groups in the area as they vowed to defend Boquist and the Republican senators.

For many Republicans it is now acceptable to say climate change exists, but it is not acceptable to indicate its causes are human-made. The problem for many Republicans is that climate change highlights the harmful consequences of industrial capitalism. In denying the problems of climate change facing humankind, conservatives defend the industrial capitalist order.[68] Believing in climate change would admit that industries and powerful businesses, along with growth-oriented economic systems, had something to do with the crisis. It would also suggest that industries would need to be regulated differently, which for corporate Republicans would open up a Pandora's box. Any mitigation-type policy or proposal supporting fossil fuel reduction or the reduction of harmful emissions is seen as deeply threatening to powerful political and economic actors. It is important to remember that many of the wealthiest and powerful families in the United States received their fortunes within the fossil fuel industry and that a very large portion of manufacturing jobs within the United States are connected with the use of fossil fuels.[69]

Addressing climate change also goes against the conservative/libertarian idea that the government is only bad and shouldn't be used to address problems that can negatively impact industry. They believe in the myth that it's best not to touch the market and that everything will work out in everyone's best interest. The corporate-funded conservative think tanks and the fossil fuel corporations have similar interests in trying to convince the US public to oppose any sort of government regulation. However, even more so, the message that government regulation is bad and that government is bad in general must be placed into the minds of the American people in order to draw a powerful line in the sand against any future where significant regulation of businesses in favor of public interests could exist.

While not all Republicans support this perspective and oppose addressing climate change through legislation (for example, Senator John McCain actively supported some legislation), most do, as conservatives are less likely than Democrats to report beliefs that are consistent with the scientific consensus on climate change.[70] Conservatives also receive more money from the fossil fuel industry. It spent $84 million on congressional campaigns in 2018, rewarding US legislators when they opposed environmental protections. About 88 percent of the oil and gas industry's campaign donations went to Republicans that year.[71]

Indeed, conservatives are more likely to defend the dominant social paradigm and the Manifest Destiny about humans and their interaction with nature. These paradigms constitute a worldview in which modern societies are seen as controlling nature, which make it difficult to accept that global warming poses any sort of significant threat.[72] Historically entrenched preoccupations with free markets and economic growth, combined with the asymmetrical political influence of powerful interest groups backed by carbon industries, have also influenced denialism, which has strengthened scientific skepticism about anthropogenic climate change and environmental decline.[73] An adherence to free market fundamentalism, which posits that the market is the only legitimate allocator of resources, is a clear barrier in the struggle to address the climate emergency.

Libertarianism, which emphasizes individual freedom while opposing state regulation, also influences Republicans and some Democrats.[74] Conservatives tend to focus on personal intentions instead of on causal mechanisms of global warming. Many adhere to the myth that a liberal elite is the primary sponsor and beneficiary of efforts to regulate the environment.[75] For many conservatives and some Democrats, disbelieving climate change and thus rejecting changes to the fossil-fuel-based economy becomes a convenient way of dealing with challenges to worldviews. In fact, scientists call this "identity-protective cognition," where people deflect threats to the identities they hold by ignoring information that is nonsalient to their perspectives.[76]

Democrats

While the forefront of attacks against addressing global warming include people mostly within the Republican Party, this does not mean that the

Democratic Party is the savior of the planet. The Democratic Party is not doing enough to address the crisis and will be hesitant to make any significant institutional changes because they would negatively impact their corporate donors, a group they have cozied up to since Bill Clinton's presidency. There is also a false belief among Democrats that a few market regulations of minimal impact will be able to fix the climate crisis. This "politically convenient" approach is less likely to challenge their corporate funders in industry. The keystone of the Democratic Party's efforts to address climate change, otherwise known as President Obama's Clean Power Plan, did not go far enough and did not approach the IPCC's target for GHG reduction.[77]

In fact, the Democratic Obama administration spoke about the negative consequences of global warming, but continued to support fossil fuels projects around the world. According to the *Guardian* newspaper, "This unprecedented backing of oil, coal and gas projects is an unexpected footnote to Obama's own climate change legacy. The president has called global warming 'terrifying' and helped broker the world's first proper agreement to tackle it, yet his administration has poured money into developments that will push the planet even closer to climate disaster."[78] The Obama administration poured nearly $34 billion into 70 fossil fuel projects around the world through the US Export-Import Bank.[79]

Numerous funding arrangements from the fossil fuel industry to the Democratic Party have existed for decades. Varshini Prakash, a member of the millennial-led Sunrise Movement, has argued that corporate donations from the coal, oil, and gas industries have challenged Democrats' abilities to pursue more ambitious climate policies. Combined, the oil and gas industry, electric utilities, and pipeline companies gave nearly $20 million to federal-level Democratic candidates during the 2016 election cycle. Hillary Clinton took in $892,252, which was only $22,000 less from fossil fuel companies than Trump took in. The American Petroleum Institute contributed $700,000 for the Democrats' 2020 convention in Philadelphia, while the coal lobby, through Duke Energy, the nation's second largest carbon emitter, gave the Democrats a $6 million loan it never repaid.[80] Moreover, the centrist Democratic think tank that grew out of the party's probusiness wing, the Progressive Policy Institute, was partially funded by ExxonMobil, which has funded climate denial efforts for decades. It received $50,000 through the Third Way Foundation according to the oil giant's 2018 Worldwide Giving Report.[81]

Prakash noted that "the GOP has largely given itself over to being the party of climate denial and raked in millions of dollars from oil and gas

CEOs as a result, and Democrats have utterly failed to put forward any practical solution to decarbonize our economy and move towards a 100% renewable energy economy."[82] One would assume the Democratic party would want to further differentiate itself from the Republican party by distancing itself from the fossil fuel industry, but that doesn't appear to be the case. The opposite seems to be occurring. In August 2018, the Democratic National Committee faced significant criticism for its decision to lift a ban on fossil fuel company donations that was passed just a few months prior. Democratic National Committee chair Tom Perez indicated his reasoning was to show that the Democratic Party supported fossil fuel workers and also employers' political action committees.[83] Moreover, Perez's Democratic Party leadership also blocked a debate on climate change in the 2020 presidential campaign, even when the candidates and party members preferred for it to move forward. On top of that, the party issued a threat to lock out any candidate who participated in a climate debate sponsored by another organization.[84] Ironically, when one television network allowed a climate debate, 2020 democratic presidential candidate Joe Biden headed to a fundraiser cohosted by Western LNG, a natural gas production company, the day after.[85]

James Hansen lists many Democrats as culprits in the battle to fight the climate emergency. He names Jerry Brown, the governor of California, for pretending to solve the problem, and he has particular disdain for Barack Obama, noting he failed miserably on global warming. He accused Obama of passing up an opportunity to thwart Donald Trump's destruction of US climate action when he declined to settle a lawsuit that would have demanded a reduction in emissions. If this had been settled, Hansen noted that Trump would not have been able to do all of the things he did during his administration, like opening up new fossil fuel sources.[86]

Connecting Political Will to Media

Mass communication, through media, journalism, and other entities, is directly connected to politics, which in turn influence media content. It is also important to remember that the vast majority of US media consist of private sector profit-driven corporations. About 90 percent of the media are owned by five massive conglomerate corporations that have a number of holdings in multiple industries. Mainstream news media tend to support the status quo, and they most often focus on elites, such as politicians, corporate CEOs, or celebrities, in their reporting. This means that general media coverage, and in the specific case of this book—of global warming—can often portray the topic from their perspectives. To be succinct, mainstream media are part of the elite, and are most likely to reproduce their worldview, which generally supports dominant political and economic power relations instead of universal societal interests. This is challenging because the views of industry and the elite often clash with the reform needed to effectively address the climate emergency.

The two-party US political system and its funding mechanism further complicate the media's relationship with the climate crisis. As this chapter and others note, monied industries that fund political parties are often opposed to policies that mitigate the climate emergency, which influences party perspectives. When you have a two-party political system, where neither party is fully invested in making the appropriate changes required to address the climate emergency, their perspectives tend to impact coverage due to journalist sourcing practices.[87] This means the diversity of the political spectrum, and its funding mechanisms, indirectly influence the diversity of global warming coverage in mainstream media. This structure challenges efforts to adequately address the climate emergency and points to the need for media reform. Chapters 8 through 10 go into more detail about this relationship.

This chapter contextualized the political will to address the climate crisis. While it is clear that most of the US political class[88] is not doing enough to address the climate emergency, it also means that citizens will need to step up to address this emergency by thinking of new and creative ways to tackle global warming. The next section moves into the field of mass communication and constitutes three chapters that examine the relationship between advertising and climate change. It presents a new way to help mitigate and address the emergency. Chapter 3 looks at the underutilized history of advertising and its relationship with global warming by connecting advertising with consumerism, industrialism, and ideology. Chapter 4 finds that advertising is interdependent with the roots of global warming, which suggests that changing the advertising system can likely alter the path of global warming. Chapter 5 rethinks advertising as a mitigative model for environmental sustainability to help address the climate emergency.

II

Advertising, Climate Change, and Reform

CHAPTER THREE

Advertising, Consumerism, Industrialism, and Ideology

Advertising is not a simple phenomenon.
—John Kenneth Galbraith[1]

While the previous section laid out the urgency of the climate crisis and examined the US political will to address the climate emergency, this chapter constitutes the beginning of the second section of the book, which examines advertising, global warming, and reform. Within the entire spectrum of mass communication media and processes, advertising constitutes the field's most pronounced contribution to the climate crisis. Advertising became an important component within capitalism through its ability to promote a consumer society as the Industrial Revolution progressed. This chapter discusses the historical relationship between advertising and the burgeoning crisis. I focus on the period before the Industrial Revolution, the transition to capitalism, and how advertising became subsumed under capitalism to push forward a consumer society during industrialization. Advertising transactions and their distribution helped operationalize a perspective of dominance over nature, which became part of their ideological orientation under a capitalist mode of production. Advertising is often ideological under different systems of production, but history doesn't suggest that advertising automatically becomes a positive contributor to environmental sustainability outside of capitalism. Nonetheless, advertising helped propel the climate crisis forward, contributing the most during the later state of industrial production, where extremely elevated consumerism and commercialism driven by advertising massively increased GHG emissions.

Advertising is having profound consequences for societies around the globe even though its intent is supposedly only for affecting sales. Indeed, advertising is a central storyteller in commercial societies and influences collective consciousness. "Advertisements are a familiar phenomenon and their extent and pervasiveness make them an integral part of both the culture of society and its economy," according to scholars John Driver and Gordon Foxall.[2] Advertisements sell not only commodities but an entire way of life that requires commercial products to function. Advertising has evolved to such an extent that it has become a major force that rivals education and religion in shaping public values and aspirations.[3]

Advertisements tell us that people will need products to be smart, happy, beautiful, and popular. According to advertising scholar Richard Pollay,

> The intent of advertising, especially in the aggregate, is to preoccupy society with material concerns, seeing commercially available goods and services as the path to happiness and the solution to virtually all problems and needs. In so doing, advertising makes consumption a top-of-mind behavior. This state of mind seems natural or rational because this persuasion also provides a world view with a value scheme that rationalizes such behavior and presents itself as commonplace.[4]

In this sense, advertising creates a lifestyle and culture of excess. Moreover, Julian Simon, former University of Maryland professor, notes that "the evidence that advertising can affect people's behavior is immediate and overwhelming, both from introspection of our own experiences and from the fact that advertisers spend hard cash to pay for advertising."[5]

Many scholars have offered definitions of advertising. Some see it as "a paid, mediated form of communication from an identifiable source, designed to persuade the receiver to take some action, now or in the future."[6] Others view it as "brand-initiated communication intent on impacting people."[7] Critical scholars Max Horkheimer and Theodor Adorno view advertising as representing social power, reflecting the structure of domination and power within a given society.[8] Many economists, mass communication scholars, and nonacademics view advertising as an ubiquitous form of propaganda that seeks to persuade the receiver to take some action (like the purchase of a good or service) or accept a certain perspective (most often benefitting the sender financially).[9]

Without assigning it positive or negative agency, advertising can be understood as a medium by which ideas are diffused in society. It entails, but is not limited to, transactions, content, and distribution. It takes many forms, but posters, television segments, newspaper designs, and online images and messages are the most common. It is usually paid for, but can be free, and it is important to know that not all advertising content promotes consumption. For example, "green advertisements"[10] can promote "eco-conscious" ways of consuming, while nonprofit, nongovernmental, and educational organizations can create advertisements that question capitalism, consumerism, and the role of advertising in society. However, this type of content is relatively rare, compared to the omnipresent advertisements for commercial purposes.[11]

Historical Contexts

As this book points out, advertising is having a negative effect on the environment as it contributes to global warming. Understanding a brief history of advertising helps us better understand and situate this contribution. Before the transition to the capitalist mode of production, advertising was essentially nonexistent as consumerism and excess were largely frowned upon. In addition, carbon dioxide emissions were harmonious with the planet and represented a natural balance with the Earth. Advertising grew in a symbiotic and mutually dependent relationship with industrialism and capitalist production capabilities, and expanded as media evolved into national mass markets. As industries within capitalism became more concentrated and less competitive during the twentieth century, competition by advertising largely replaced competition by price within oligopolistic markets. As these markets became more concentrated, advertising's importance inflated further. It evolved to become a crucial mechanism within capitalism that further advanced the antagonism between ever-expanding production processes and the health of the Earth. The Industrial Revolution helped propel a 44 percent increase in carbon dioxide levels by 2015 according to the World Meteorological Organization.[12]

Before the Industrial Revolution and the Transition to Capitalism

Prior to the Industrial Revolution life in the Middle Ages was often based on religious scriptures and interpretations of commerce. These influences suggested that people conceive of their lives as transitionary periods between the "nothingness of pre-existence and the glorious eternal life in the 'heavenly city,'" according to James W. Carey, former professor of communication.[13] Humans were supposed to primarily serve God and the heavenly gains of salvation. The pursuit of wealth and the drive for material gain came in direct conflict with these goals as they were often seen as immoral. According to Carey, the "economic ethic of the medieval period is summed up in the statement: homo mercator vix aut numquam potest Deo placer ('the merchant can never or hardly ever please God')."[14]

This changed as the Renaissance and the Reformation emphasized individualism, in particular on the development and growth of individual human potentialities. This new perspective spread across Europe to eventually dominate the latter half of the eighteenth century with burgeoning commerce accompanying its growth. Capitalistic forms of economic life were beginning to take hold with the use of money, the quest for wealth, self-interest, and new market-based economies. While much of Christian teaching after the Reformation encouraged the pursuit of wealth and materialism, indulgence was to be frowned upon.[15] This transitionary period from agrarian, handicraft society to the eventual mechanized era was influenced by classical liberalism, which stressed the role of the individual above all.

While living under medieval lords was clearly not the most pleasant way of life for the majority of European inhabitants, the transition to capitalism wasn't enjoyable either. Capitalist consumption was often born out of painful and bloody events involving the removal of people from their means of autonomy. Prior to the Industrial Revolution, more than 90 percent of the US population lived on farms, where they were relatively autonomous.[16] Edward Comor, associate professor at the University of Western Ontario, notes,

> By stripping populations of their pre- or non-capitalist modes of subsistence and hallowing out traditional ways of finding identity and meaning, individuals and collectivities are compelled to change. Whether the particular circumstance of this involves the "enclosure of the commons," the "privatization of collective assets" or the transformation of norms through the barrel of a gun, the birth of capitalist consumption invariably entails some sort of violent "push."[17]

Introducing Capitalism, Advertising, and Consumerism

Capitalism is a historically specific mode of social and economic organization that is distinguished from other socioeconomic systems. It concentrates the means of production into relatively few hands, forces a substantial portion of the population to exchange their labor in return for a wage, and allows capitalists to own the products and production processes as they bring the products to markets to realize profits. Moreover, capitalism relies on a monetary system for production through bank credit and on market exchange, and its primary motivation is profit.[18]

Before the capitalist system of production went through major changes to become highly concentrated, advertising was essentially relegated to the periphery of any influence to the economy and mainly consisted of proclamations and announcements.[19] With mostly local markets in pre-twentieth-century capitalism, advertising accounted for well below 1 percent of the GDP. Early forms of advertising had been employed to support economic expansion and capitalist development, even before industrialization. One of the early uses of advertising, during the colonial era, was to persuade people to join the imperial colonization process to the "new world" by promising fortune; other scholars see its origins in the slave trade through notices for runaway slaves.[20] In both cases, the purpose of advertising was to encourage the plunder of raw materials and people from the colonies by the colonizing countries in order to establish consumerist economies. This ideological component of early advertising began with capitalism itself and its relationship with consumerism. Advertising could not ignore capitalism, as it quickly became deeply and formally subsumed under its development.[21]

Consumerism is unique to capitalism. It is a way of life that is dependent on capitalism. Marx noted that humans have always produced and consumed, but *how* they go about doing these things is always historically situated.[22] A consumer society is the outcome of capitalism's need to sell what it produces, and consumerism is the ideology that seeks to match consumption with production.[23] As capitalism developed, and productive capacity expanded, it required consumers, so the public's sense of need for goods had to grow in tandem. The crucial role of advertising and consumerism within this landscape is to construct needs and demands, consisting of promises of better lives and happiness.[24]

In essence, the mode of production influences the types and goals of advertising that each society produces, or, in other words, "society lays the foundations from which advertising as a whole draws its lifeblood," according to economists Paul Baran and Paul Sweezy.[25] This suggests that capitalism, socialism, and other modes of production would each create different and unique systems for advertising. Advertising in capitalism, which is its dominant context, is extremely complex and unique in that it exists as both a dependent and independent variable within capitalism. Capitalism generally entails commercialized, market-based media structures, where individuals or organizations pay to place messages within media through advertising. Advertising exists as a dependent variable in that it is dependent on capitalism for its standard mode of operation and main identity. The profit-centered goal of capitalism largely defines and determines the distributive parameters of advertising transactions by limiting them as business exchanges devoid of any larger social or environmental concern.

In addition, advertising also exists as a dependent variable because it is dependent upon content from outside influences (society). Content, besides distribution, as another part of the advertising process, is dependent on outside influences, meaning NGOs, corporations, or governments that create a wide range of content. Therefore, advertising is neither completely autonomous nor independent in this regard. However, advertising should also be considered an independent variable in that it can influence changes to other variables within capitalism, such as consumption levels, economic growth, and indirectly to the environment. For example, it can stimulate and grow the economy through its potential to increase sales of items, while it can also encourage overconsumption of natural resources, which in turn harms the environment.[26]

Advertising and US Industrialization

Advertising grew in a mutually dependent and symbiotic relationship with each wave of industrialism and capitalist production capabilities. It burgeoned as the first wave of industrialization created steam engines and new manufacturing facilities, which resulted in a need for more consumers. As the media developed in the United States, Volney Palmer, who claimed to represent 1,300 newspapers in the 1840s, emphasized to his clients the critical importance of advertising as a means to develop and expand new business markets.[27] Advertising revenue represented only 44 percent of newspaper income in 1880, but by 1900 it had increased to 55 percent.[28] However, not everyone was excited about new forms of consumption being pushed by advertising. During the 1890s, department stores in Chicago were referred to as "inventions of the devil," "producers of crime, sorrow, and disgrace," "monsters," and "vultures" by the North Side Businessman's Association.[29]

Electrically powered production ushered in a second phase of industrialization during the late 1800s to the early 1900s, which further divided labor and coincided with the flooding of a more diverse array of products into markets, thanks to advertising. In 1790, 80 percent of all clothing was made in the home for family use, but by 1890, 90 percent of boys and men's clothing was made outside the home, while women's clothes were made according to patterns purchased from women's magazine companies.[30] During this period advertising increased tenfold.[31] It grew from a $200 million business in 1880 to a $3.5 billion business just before the stock market crash of 1929.[32] The early 1900s also saw the development of scientific methods to gauge consumer opinion, while business schools began to teach strategies and tactics of advertising. The Associated Advertising Clubs of America formed in 1905, and by 1916 had more than 15,000 members.[33] Following World War I, advertising continued to grow significantly with businesses doubling the amount of money spent on advertising their products.

The evolving media, which extended the consumer base through new mass markets, greatly affected the growth and influence of adverting in society. In particular, the national reach of the new commercial broadcast media system, starting around the 1930s, led to a massive expansion of advertising. In essence, advertising would catapult and drive media expansion, while cultivating a desire for consumer products, now at the national level, which would further expand productive capacities for manufacturers.

After the Depression, during the third stage of the industrial revolution, new technologies, computers, telecommunications, and expanded automation further increased productive output, creating a major manufacturing base. Advertising saw another period of growth after World War II, partially thanks to national television mass markets. The annual average business investment in advertising between 1950 and 2003 increased dramatically. In 1950–59 business investment in advertising was $8.6 billion. In the 1960s it increased to $15 billion, in the 1970s to $30 billion, in the '80s to $89 billion, in the '90s to $165 billion, and by 2003 it was at $240 billion dollars.[34] Another estimate from 1980 and 2003 put real dollar advertising expenditures increasing from $135 billion to $231 billion.[35]

The time period between the 1970s and the 1990s also saw a tenfold increase in the number of new commodities joining the market.[36] To take one year for example, in 1990, there were 12,055 new products, or 33 per day, introduced into American drugstores and supermarkets. Many of these products were often indistinguishable or slightly different from existing products except for their packaging and only about 10 percent were actually new to consumers.[37] To accommodate these products, there was also a major increase in the size of retail space, especially throughout the United States.[38]

Even though criticism of advertising increased during this time, painting unflattering pictures of admen as greedy con men or hucksters, advertising eventually won the power struggle, and from the 1960s onward began to regularly bombard consumers with messages from every direction, at just about any time. Twenty years after that, the average American adult was exposed to about 3,000 ads per day.[39] Today, some studies suggest the average North American is now exposed to up to 10,000 marketing messages a day.[40] These numbers are likely to increase in the future as advertising around the world has been growing steadily since 2009.[41] The amount of money spent annually on advertising in the United States exceeds total US federal, state, and local government public expenditures on police protection, natural resources, and higher education combined.[42]

By 2006, that translated to roughly $280 billion spent on advertising expenditures in the United States, where there are more than 33,000 nationally advertised brands.[43] US digital advertising totaled $51 billion in 2014, which was more than was going to all print media, and by 2018 was expected to increase to $82 billion annually.[44] Search advertising constituted about half of online advertising around that time, which made Google extremely wealthy.[45] According to the World Federation of Advertisers, which represents about 90 percent of global marketing communications through 55 national advertiser associations on five continents, roughly $700 billion per year is spent on advertising that tries to persuade potential consumers to purchase more and more.[46] Global advertising media owner revenues were expected to reach $536 billion in 2015, while the United States was expected to pull in $165 billion.[47] In fact, global media owners' net ad sales rose by 5.7 percent in 2016.[48] For the United States, which is considered one of the world's largest marketplaces, its retail sales topped $4.53 trillion in 2013, which led economic growth representing 27 percent of nominal US GDP.[49]

Not surprisingly, with increases in advertisements, combined with the transition to neoliberal economics pushed by Thatcher and Reagan, US household debt also skyrocketed during the last several decades, marking this hypercommercial period as quite shocking. Data from the US Federal Reserve show total US household debt between 1950 and 2006 going from virtually nothing in 1950 to over $12 billion by 2006.[50] US personal savings as a percentage of disposable income since the 1980s also decreased to become negative around 2006.[51] Economist John Kenneth Galbraith viewed advertising and social emulation as the drivers of this debt, as the need to persuade people to incur debt along with the arguments and abilities allowing them to do so were considered part of the overall production process.[52]

Beside increases in debt, the advertising-driven industrial expansion also coincided with the highest historic contributions to GHG and carbon emissions. As many societies turned into full-fledged advertising-driven consumer cultures, which evolved after World War II but solidified during the 1970s, total GHG and carbon dioxide emissions increased significantly during this period. Data from the IPCC indicate that about half of cumulative anthropogenic carbon dioxide emissions between 1750 and 2010 have occurred in the last 40 years.[53] This means that the height of our commercialized, advertiser-driven consumer societies created the largest increase in GHG emissions over the past few decades. These are significant GHG emissions during an extremely short period of time.

Additional data show total annual anthropogenic GHG emissions increased significantly from 1970 through 2010 with larger decadal increases toward the end of this period.[54] They increased from 27 to 49 gigatons (or gigaton carbon dioxide equivalents) during this time frame, with carbon dioxide fossil fuel and industrial processes as the main contributors. Their contribution increased from 55 percent in 1970 to 62 percent by 2010. Fossil fuel combustion and industrial processes contributed about 78 percent of the total GHG emission increase from 1970 to 2010.[55] Industry and fossil fuels appear to be the main drivers of the increase in GHG emissions. The significant increase in GHG emissions between 2000 and 2010 came from energy supply (47%), industry (30%), transport (11%), and buildings (3%). By 2010, the transport sector accounted for 27 percent of final energy use in 2010, with baseline carbon dioxide emissions projected to double by 2050. The same year, industry accounted for 28 percent of final energy use, with emissions projected to increase by 50–150 percent by 2050. It is important to note these increases have been happening *despite* a number of climate change mitigation processes!

The purpose of showing the extremely important relationship between advertising and industrialization is to highlight its importance in the overall process that rapidly contributed to global warming. Chapter 4 goes into more detail about the relationship between advertising, production, consumption, growth economics, and climate crisis.

Advertising and Ideology

Advertising, as a complex system that entails transactions, content, and distribution, is considered ideological under capitalism for at least three reasons. First, it is a necessary component of a consumer society. Second, it has a tendency to treat the public as consumers instead of citizens, and third, because it became subsumed under a capitalist relationship with nature.

The ideological nature of advertising within capitalism begins with capitalism itself and its relationship with consumerism. Advertising is an indispensable tool of capitalism, both practically and ideologically, as it uses the advertising industry to help organize and stimulate consumption.[56] In this sense, advertising comports with additional definitions of ideology.[57] It conveys and promotes a materialist belief system by mostly presenting the products and ideologies of powerful companies in hopes of affecting consumption, which enriches the producers of commodities (or the class that owns the means of production), in the face of opposition from groups that oppose their unsustainable or harmful production processes. Advertising can also discriminate against competing products and less powerful actors, especially noncommercial and nonprofit organizations. It also distorts the reality of harmful production processes by omission, helps legitimize sales effort processes, and enriches the ownership class with the productive capacity. Predominantly aimed at wealthier consumers, advertising ultimately aims to manipulate humans' desires, tastes, purchases, and consumption decisions. John Storey, emeritus professor at the University of Sunderland, UK, noted that a "consumer society could just as easily be called an advertising society."[58] In this sense, advertising exists as a core function within a consumer society.

The second reason why advertising is considered ideological within capitalism is because it seeks to define the public as passive consumers instead of active citizens.[59] The vast majority of advertisements under capitalism promote items and services to sell instead of presenting empowering, educational, moral, political, or ethical messages of use to the public.[60] This effectively limits and marginalizes discussions of the public good. Treating citizens as consumers evolved to penetrate media through their depictions.[61]

The commercial media are also supported by advertisers, which means they are largely funded by interests that promote consumption. From a political and economic perspective, media industries are often considered a "cultural apparatus," whose mission is to largely serve the interests of capital.[62] This apparatus is more concerned with expanding by incorporating more areas into markets and profits than with promoting less wasteful and consumptive behavior.[63] In this sense, advertising-supported media largely perpetuate a 'business as usual' approach to consumption, where people are to shop their way out of economic recessions, even in the midst of a climate crisis fueled by overconsumption.[64] Chapter 10 expands on this within the context of journalism.

Third, advertising is considered ideological because it reflects a capitalist relationship with nature. As previously mentioned, before the transition to the capitalist mode of production, advertising was minimal to nonexistent as consumerism and excess were largely frowned upon. In addition, carbon dioxide emissions were harmonious with the planet and represented a natural balance with the Earth. As advertising grew and developed it became subsumed under a capitalist relationship with nature. This suggests that the manner in which capitalism viewed and interacted with nature transferred to the way in which advertising interacted with nature. In the case of capitalist relations, advertising is generally defined in business terms as opposed to anything with societal or environmental ramifications.

This means that the present system of advertising in commercialized, market-based media structures, where individuals or organizations pay to place messages within media, is largely envisioned and conceived as a business transaction. It is a monolithic system where advertising is limited as a commercial endeavor. This "bias" means that the transactions and distribution within advertising processes are oblivious to important consequences of their mode of operation. An advertising transaction occurs when an individual or organization arranges, usually by purchasing space, to have an advertisement appear in a medium of communication. Distributive aspects of advertising occur when the owners of the communication medium decide to publish and distribute the advertisement to an audience. The transactional and distributive aspects of other industries—like alcohol, tobacco, explosives, weapons, medicines, certain kinds of gasses, chemicals, within restaurants, and many other areas—are significantly regulated because they can cause harm. These transactional and distributive aspects of other industries do not apply to advertising, even though advertising also negatively contributes to environmental and human health.

This suggests that advertising in general mostly excludes both the pre- and post-advertising stages of products or services that are advertised, while the transactional and distributive aspects are completely blind to them. Even with its power over influencing consumption,[65] the transactional and distributive aspects of advertising ignore the production processes of the products or services to be advertised and usually discount the human and environmental consequences of their being advertised through various media. These aspects are underexamined in establishment advertising literature, as the preference is to focus on advertising content or consumers.

This operationalization of advertising and its complete rejection of environmental concerns within its transactional and distributive processes stems from what Adorno claimed to be capitalism's placement of nature behind capital's economic dimensions in order to ensure their survival within capitalism.[66] This perspective exhibits a domination aspect in that for material progress, the creation of liberal individualism, individual subjugation, and domination over nature became necessary components.[67] It branches from Renaissance and Reformation emphases on the growth of individual human potentialities accompanied by expanding commerce and its growth. Only when capitalism established itself did the domination over nature in the interest of the entire economic system paradigm take over.[68]

Ideology outside Capitalism

The ideological nature of advertising is not limited to capitalism and exists in other economic systems. One might assume ideological relationships outside of capitalism to be less exploitative for humans and nature. However, as long as the transactional and distributive aspects of advertising exchanges ignore the relationship between production processes and nature, this dimension could exist in any number of political and economic systems. Advertising may not exhibit a healthy relationship between production processes and nature just because it exists in socialism. In other words, every product or service advertised in socialism or other system outside of capitalism doesn't guarantee that the production process of the product or service advertised isn't exploitative and harmful to nature.

Among those who examine socialism and ecology, theorizing advertising's relationship with nature comprises little if any space at all.[69] This is most likely because there is practically no advertising in a classical socialist economy.[70] Advertising in a socialist economy would be relatively unimportant and unnecessary from a social standpoint.[71] Nonetheless, within the many different types of socialism, commercial advertising existed and continues to exist in socialist countries.[72] Burnham Beckwith suggested that wherever advertising existed it should be complemented with equal space for secondhand goods in order for the public to compare each item.[73]

While this is a start, eco-socialist conceptualizations of advertising do not appear to be formally developed as of yet.

This means that without a defined socialist theory of advertising and ecology, advertising in those systems may also ignore the ecological ramifications of its usage. Naomi Klein, award-winning author and Gloria Steinem chair in media, culture, and feminist studies at Rutgers University, provides an insightful discussion on this topic:

> We have to be honest that autocratic industrial socialism has also been a disaster for the environment, as evidenced most dramatically by the fact that carbon emissions briefly plummeted when the economies of the former Soviet Union collapsed in the early 1990s. . . . Let's acknowledge this fact, while also pointing out that countries with a strong democratic socialist tradition—like Denmark, Sweden, and Uruguay—have some of the most visionary environmental policies in the world. From this we can conclude that socialism isn't necessarily ecological, but that a new form of democratic eco-socialism, with the humility to learn from Indigenous teachings about the duties to future generations and the interconnection of all of life, appears to be humanity's best shot at collective survival.[74]

Klein is not alone believing that neither state socialism, variants of capitalism, nor the emerging green Keynesianism offer convincing answers to the existential crisis.[75] James O'Connor, former professor of sociology and economics at the University of Santa Cruz, California, finds that centralized socialist countries are likely to have used up nonrenewable sources, as well as have polluted their environments as fast or faster than their capitalist counterparts.[76] Part of this comes from "catching up with the West," as well as resistance to Western aggression in the context of the Cold War. Nonetheless, he points out that socialist revolutions have been far less ecologically harmful than capitalist counterrevolution. Socialist economies stress collective consumption and waste fewer resources than the individual consumption trends in capitalist economies. Moreover, much of the world's ecological harm has been caused by imperial capitalist wars combined with the Global North's imperial mode of living.[77]

Regardless of the economic system, there is significant debate about whether economic growth is even compatible with a sustainable planet. Much of socialist thought has been premised on the continuity of material trajectory from capitalism to socialism,[78] while growth targeting has consistently been a driver of socialist plans and social-democracy programs.[79] Degrowth scholar Giorgios Kallis argues that in addition to capitalist growth, "any type of socialism that pursues and produces growth will be unsustainable."[80] Others believe that the environmental destruction caused by capitalism is not due to growth, but to capitalist accumulation.[81] In effect, by abolishing elements that drive capitalist accumulation, growth could be greened. On the other side of the debate, Kallis believes a sustainable socialism would need to plan for declining material and energy use, which means decreasing industrial output.[82]

These perspectives clearly demonstrate there is a significant need to develop new conceptualizations of advertising given its prominence within economic systems around the world. One must ask if it's possible to change the manner in which advertising operates to support the consumption of mostly renewable resources that create less waste in nature. Perhaps eliminating capitalism's relationship within advertising could help improve mitigation efforts? Would it be possible to alter advertising in order to promote primarily Earth-friendly sustainable products under capitalism? Could a new advertising system contribute and help transition out of the capitalist mode of production if that is one of main causes of the global warming crisis?

The next chapter addresses similar questions as it examines how advertising is interdependent with the main influences of global warming. It looks at how advertising is related to production, consumption, growth economics, and the global warming crisis. Connecting these mechanisms while noting they are interdependent suggests that altering the advertising system would also change the current trajectory of the crisis.

CHAPTER FOUR

Advertising and Its Interdependence with the Origins of the Climate Crisis

The previous chapter examined advertising during the transition from a preindustrial society to a fully industrial and capitalist, consumer society. In this chapter I examine relationships between the dominant human-made mechanisms that form the roots of global warming to develop a materialist understanding of the relationship between advertising and the climate crisis. The IPCC views the production and consumption of goods and services, aided by advertising, as key drivers of global warming through their contribution to GHG emissions.[1] These mechanisms exist within a historical global political and economic structure rooted in growth economic policies. Understanding the crisis within this context helps us understand the power dynamics that influence global warming, and how advertising interacts with the main mechanisms of human-made climate change: production, consumption, and growth economics.[2] With limited time to reverse the trajectory of the climate emergency, perhaps this knowledge can inform new approaches and policies to assist civic groups in reversing the crisis's trajectory.

I make two arguments in this chapter. First is that the manner in which advertising interacts with production, consumption, and growth economics suggests these mechanisms are *interdependent* in their ability to influence global warming. Arguing that they are interdependent is relatively new, which has significant implications. This interdependence implies that significantly altering any one of the four mechanisms would likely change the present trajectory of the climate crisis. The second argument, which is premised on the first, is that there is a need for scholars and civic groups to reassess the role of power relations and agency within advertising.[3] Its interdependence with the other mechanisms suggests that if modified, it has substantial aptitude to alter these mechanisms, and thus positively impact global warming. Therefore, civic, activist, and political groups interested in mitigating the climate crisis may want to consider changing the advertising system as an additional strategy to address climate change.

This chapter begins by defining and tracing the relationships between advertising and the other main human-made mechanisms that influence global warming. The trajectory of how these relationships negatively affect the environment is then discussed. Scientists connect them to a number of negative environmental externalities stemming from profit-driven production processes along with imperatives for unending economic growth. Inequalities of consumption at both the local and global levels are examined in terms of their contribution to environmental degradation, while the levels and locations of GHG emissions that support unequal levels of consumption are also explored. Last, I connect these trends to suggest advertising can have a purpose in the struggle for a sustainable future.

Tracing the Relationships between Production, Consumption, and Advertising

Production is generally understood as a process that combines immaterial inputs, like specific plans, with material inputs, in order to make something for consumption or output, like a good or a service with a specific value.[4] There are at least three different kinds of production processes: market, public, and household production. Within economics, a market is generally any structure where buyers and sellers exchange goods, information, or services. In this chapter I focus on market production, as it generates the bulk of all advertised services and manufactured commodities.

Consumption, or consumption expenditure in economic terms, in its simplest form is the final purchase of goods or services by individuals. Viewed by economists as the final purpose of economic activity, consumption, combined with production, are viewed as two of the most important variables defining economic performance. According to economist Chen Dongling, "consumption expenditure constitutes the largest individual share of gross domestic product in most countries around the world."[5]

As we saw in chapter 3, the Industrial Revolution significantly enhanced commodity production output, which preceded and drove consumption.[6]

Suddenly, selling the massive increase in manufactured items became a major challenge.[7] Economist James Kenneth Galbraith viewed production as creating more wants and then the need for more production.[8] He called this dynamic the "dependence effect." If production was to be increased, then more wants had to be generated. He found the "higher level of production has, merely, a higher level of want creation necessitating a higher level of want satisfaction."[9] In this model, production mediates consumption, while consumption also mediates production. Without production there is no consumption; without consumption, no production.[10]

Introducing Advertising with Production and Consumption

For capitalism, the purpose of advertising was to create demand in order to match consumption to production.[11] To address the challenge of selling the large quantities of manufactured items from the Industrial Revolution, advertising became a necessary component to encourage consumption by creating demand and new markets.[12] Much of production in capitalism has little to do with the satisfaction of genuine human needs and more to do with the profit motive that drives the extraction of raw materials to produce a never-ending supply of goods.[13] Without advertising, potential consumers would be less informed about items for purchase. For advanced capitalist societies, individual dispositions to consume must be socialized, as consumerism must drive consumption.[14] This means advertising encourages materialistic values to further even more production and consumption.[15]

According to one well-known framework, advertising is seen as an explicit good in consumers' utility functions and affects product demand due to complementarities in the joint consumption of advertising and products. In other words, consumers who consume more advertising have a higher value for the advertised product. Here the consumer is in a feedback loop, so consuming products and consuming advertisements have a similar influence on the overall consumption experience.[16] At this point more consumption of an ad induces more consumption of the product, and more consumption of the product stimulates more consumption of the ad.[17]

Obviously advertising does not work all of the time and exposure to advertisements does not assume causality for consumption. However, over time, the macro effect of advertising suggests it is an extremely important variable in terms of influencing consumption.[18] Scholars Robert Brulle and Lindsay Young indicate that advertising "plays a significant role in the expansion of personal consumption,"[19] while researcher Erik Assadourian notes "if advertising were not effective, businesses would not spend 1 percent of the gross world product to sell their wares, as they do."[20] With the goal of infinite surplus generation, businesses continue to seek growth and additional profits generated by new products and services, more wants and further consumption.

Growth Economics and Advertising

Economic growth as a policy goal appeared in the 1950s and is traced to the United States, Britain, Japan, and other countries.[21] The growth paradigm is largely accepted and unquestioned in advanced and developing countries.[22] Growth economics demands that an economy continually increase its capacity to produce goods and services.[23] According to award-winning economist Simon Kuznets, "a country's economic growth may be defined as a long-term rise in capacity to supply increasingly diverse economic goods to its population."[24] In this sense, commodification is a driver of economic growth.[25]

Although often gauged by examining productivity and output, growth economics can also include consumption, as production and consumption contribute to defining performance. According to Edward Comor, the entire structure of the economy is built on a consumption-dependent framework, as consumption exists as a dominant force that drives capitalist economies and economic growth.[26] Gross National Product (GNP) or Gross Domestic Product (GDP) are economic indicators that measure economic performance by focusing on the total amount of money being exchanged at one point in time. This means that in many places around the world, and especially in the United States, people are encouraged to believe that enhanced consumption is necessary to sustain prosperity, and even that consumption is a patriotic duty.[27]

Echoes of this can be found in President Dwight D. Eisenhower's suggestion that Americans needed to help buy the nation out of the recession during his presidency.[28] Ironically, decades later, the same argument would rise again during the George W. Bush presidency when he informed Americans of the "need" to go shopping during an economic lull after the September 11, 2001 attacks. These exhortations situate consumption as not only part of economic growth but also of national purpose. Trish Morgan, lecturer at Dublin City University, notes that "just as shopping was the most patriotic act an American could have done post 9/11, in the phase of monopoly-finance capital the indebted citizen is once again charged with driving the economy back to growth while placing unprecedented stress on the finite resources of the ecosystem."[29]

Advertising fits into the goal of economic growth, as it was considered part of the economic development mechanism,[30] which economist Joseph Schumpeter noted included new ways of commercially handling commodities.[31] It does this by enhancing a consumer-oriented society.[32] Paul Baran and Paul Sweezy also found that advertising could spur economic growth by stimulating consumption, spending, increased production, and investment opportunities.[33] It became a critical sector of the total economy, which maintains and perpetuates political and economic control,[34] and is thus necessary to growth/consumer-oriented economic systems and profit-oriented commodity production processes. Advertising evolved to become inseparable from the economics of consumption and became an integral part of the logic by which commerce and growth are moved forward.

In essence, "promotion surely has a stimulating effect on the development of new consumer-durable industries. Without promotion the inventor and entrepreneur could not count on rapidly expanding the market," according to professor Julian Simon.[35] Moreover, there "is a strong relationship between the extent of advertising in an economy and the level of development of the economy."[36] Although some countries may deviate from this pattern, it is quite strong and clear—generally the richer the country, the larger the proportion of money spent on advertising.

Advertising, Economic Growth, and Consumption

Research during the mid-twentieth century, which saw a massive increase in productive capacity, corroborated earlier studies that attributed to advertising an ability to influence economic growth and consumption. One study during the great push to a consumer society in the 1960s found that

> an additional increase of 1% in advertising expenditures beyond and above the rate of increase in Gross National Product can generate one-tenth of one percent increase in consumption. In absolute figures, an additional amount in $150 million in advertising expenditures can bring in about an increase of about $400 million in consumption.[37]

Overall advertising expenditures (as defined by McCann Erickson, the Internal Revenue Service, and *Advertising Age*) increased at the same time as Gross National Product and personal consumption expenditures increased during the 1930s through the 1960s.[38]

As one industry guru proposed, "Business men and economists have few doubts that advertising increases consumption for the economy as a whole. In fact, the opinion of our profession seems almost unanimous."[39] In essence, advertising affects the propensity to consume at the same time as it affects consumer choices among brands.[40] With economic development and performance directly connected to levels of consumption,[41] advertising became a necessary component of market systems[42] that would move overall consumption forward by creating messages about dissatisfaction, the "need" to overconsume, individualism, and materialism.[43] Overall, advertising should not be separated from the economics of consumption or the Gross Domestic Product.

However, economists like Galbraith were hesitant to give advertising significant agency as exchange can happen without advertising, and exchange is how economic growth is measured.[44] Simon expands on this by noting that exchange can exist without promotion, but promotion cannot exist without exchange.[45] Given the relative newness of modern mass advertising and the lack of an adequate theory at the time to fully explain how advertising functioned in society, Galbraith and other economists focused on production more than advertising. In addition, advertising may not have been considered a fully developed academic field of inquiry at the time of Galbraith's writing. Galbraith was uneasy about connecting advertising to production and demand largely because "it cannot be accommodated easily to existing theory."[46] Beyond that, "there has been little tendency to go on to examine the implications of this for the theory of consumer demand and even less for the importance of production and productive efficiency."[47]

In the development of hypercommercial societies, where advertising became even more ubiquitous, for some scholars pinpointing the ability of advertising to impact economic development remained challenging. As economist Charles Kindleberger noted long ago, too little attention was generally paid to the spread of communication needed to link markets.[48] Surprisingly this remains an understudied area, even though its impact beyond mere economic growth is significant. This is unfortunate as some scholars clearly indicated advertising exhibited significant power, yet many weren't interested in exploring how that power could be used to explore more positive relationships between society and nature.

Human-Made Mechanisms and Environmental Externalities

The transition into capitalism and its perpetual growth paradigm is considered a significant driver of environmental problems. GDP allowed capitalist economies to assess growth indefinitely and without assessing any negative environmental externalities.[49] Moreover, there is no case of economic growth that has not involved accumulation and exploitation of nature and workers.[50] Economic growth in advanced capitalism consists of capitalist overproduction and overconsumption, which result from accumulation-driven economic processes based on the production, capture, and mobilization of a social surplus through exploitation.[51] This activity consists of two sides, with one process that appropriates and manipulates nature while the second consists of the social relations that enable the material appropriation and manipulation to be carried out.[52] These fundamental mechanisms within capitalism are problematic because they are dependent on low entropy fossil energy, while they are incapable of sustainably managing waste.[53] Data indicate that 1 percent growth of a national economy is associated with a 0.6 percent to 0.8 percent increase in carbon emissions and a 0.8 percent growth in its resource use.[54]

This suggests that one of capitalism's contradictions is that it undermines the biophysical conditions on which it depends during its never-ending pursuit of capital accumulation, which is rooted in and dependent upon unsustainable consumption of nature.[55] In this sense, capitalism sees nature as its never-ending gas station and department store, while the never-ending consumption boom harms the environment to the extent that it becomes a significant crisis for capitalism.[56] Moreover, besides being environmentally unsustainable, capitalist economic growth is also unjust.[57]

As part of capitalism, there is no doubt that the interaction between advertising, production, growth economics, and consumption creates significant environmental externalities. Scholars and scientists connect production within consumer-oriented economic systems with climate change, the destruction of the ozone layer, species extinction, the loss of fresh water and land cover, chemical pollution, and the disruption of various natural cycles.[58]

Because the Earth's ecosystem evolves but does not grow, the drive for unlimited expansion in growth-oriented economies creates an impossible and unsustainable situation, which is likely to destroy the planet.[59] This means that nonstop economic growth and profit accumulation are ultimately at odds with planetary sustainability.[60] Infinite accumulation of capital within consumer-oriented economic systems promotes consumption beyond human needs, which simultaneously ignores the limits of nonrenewable energy resources and challenges the Earth's waste capacity.[61] Gar Alperovitz, professor of political economy at the University of Maryland, finds the planet is approaching real limits to the availability of numerous basic resources necessary for economic advancement and that no technological fix is going to be able to address this challenge.[62]

The job of advertising has always been to promote profit-driven production and the imperative for economic growth under expanding industrialization processes, which generate significant negative environmental consequences. Labor and the costs of producing products from materials that harm the planet are often more affordable for manufacturers, compared to Earth-friendly and sustainable production processes.[63] Purposely designing products that break down or expire relatively quickly, otherwise referred to as planned obsolescence, is more profitable than producing long-lasting goods.[64] Galbraith believed that production models like these and the overall increase in production associated with the industrial revolution would create surplus waste and environmental devastation.[65] Moreover, the transportation required to distribute the items produced within this system also required additional natural resources for distribution. When factories began to churn out products in quantities too great to be consumed in local areas, as Naomi Klein notes, the manufacturing and global transport of these advertised goods expended unprecedented amounts of GHG emissions and fossil fuels, which are some of the primary causes of global warming.[66]

Consumption

Indeed, the elevated levels of consumption encouraged by advertising are harmful to the environment. The global economy has become increasingly dependent on consuming goods and services, even though the depletion of natural resources along with environmental degradation is directly related to the consumption of these goods and services.[67] According to scholar Steven Shapiro, "Advertising and marketing are front and center in our personal and planet-harming over-consumptive, materialistic, and self-centered ways."[68]

Individual consumption within growth/consumer-oriented economic systems is responsible for the majority of global emissions.[69] This means that the consumer system that produces goods and services (indirect emissions) contributes more emissions than do individual direct emissions (for example, the emissions produced by driving a car, or home gas use).[70] The overall consumer culture, meaning the costs of the production, use, and disposal of commodities has led to the degradation of over 60 percent of the world's ecosystems.[71] These rates of consumption are ultimately unsustainable,[72] which suggests that the consumer system is one the most important factors in global warming.

Global Inequities of Consumption and GHG Production

The levels and locations of global consumption mostly correlate with the levels and locations of GHG emissions. While mitigating global warming is a planetary issue, much of the responsibility is blamed on Western, overdeveloped countries and large industries, as they are some of the first and largest producers of GHG emissions. For example, the United States and Western Europe constitute just over 10 percent of the world's population, yet consume 60 percent of pretty much everything produced.[73] Moreover, about 90 companies, mostly in the oil, gas, and coal industries, created roughly two-thirds of cumulative global emissions of industrial carbon dioxide and methane between 1751 and 2010.[74]

The largest current emitters in terms of countries/regions are the United States and China followed by the European Union. According to a study by OXFAM, the world's richest 10 percent of people, with the vast majority living in the 34 OECD (Organization for Economic Cooperation and Development) countries like the United States, Australia, Canada, France, Germany, Israel, the United Kingdom, and Spain, produce half of carbon emissions, with a third of these emissions from the United States. OXFAM notes that individual consumption is responsible for 64 percent of global emissions, while the remaining 36 percent are attributed to consumption by governments through investments in infrastructure and international transport.[75]

According to IPCC data, high-income countries like the United States, the European Union, Japan, Russia, and Canada have 18 percent of the world's population, yet contribute the highest GHG emissions on the planet per capita. The United States, China, and the European Union, constituting almost half of the world's total emissions, exhibit the highest *total* historic emissions, or cumulative emissions. From 1850 to 2011, the United States, the EU, China, the Russian Federation, and Japan together contributed two-thirds of the world's historic carbon emissions. Of these historic emissions, the United States and the EU contributed over half of all emissions. Excluding land use change and forestry, China was the number-one producer of GHG emissions, while the United States and the EU come in second and third as the top three as of 2011.[76] However, Canada and the United States have the highest *per capita* emissions of GHGs, while India barely shows up on the scale.[77]

Middle-income countries like China, India, Indonesia, Brazil, and Mexico, where more than 70 percent of the world's population lives, create about 54 percent of global GHG emissions.[78] The so-called "less-developed" countries contribute much less, with half of the world's population essentially contributing no carbon emissions.[79] Ironically, with a shift of global manufacturing, wealthy countries are also outsourcing much of their manufacturing, and thus carbon emissions, to poor and emerging economies.[80]

US Consumption and Emissions

It is important to highlight the consumption practices of the United States, as this country constitutes one of the world's largest marketplaces and devours about 32 percent of total consumption, yet only constitutes 5 percent of the Earth's population.[81] It also produces about 25 percent of worldwide emissions.[82] Likewise, it often leads the world in the amount of materials consumed per capita and generates the highest total lifestyle consumption emissions within G20 countries, given the available data.[83] According to a study in *Advertising Age*, the top 10 percent of American households account for nearly half of all consumer spending, while the carbon footprint for the top quintuple of the US economy is estimated to be more than three times that of the bottom quintuple.[84]

There is also a trend that shows the average US resident increasing, not decreasing, their consumption during the past several decades. In the 1960s, the average American used 366 pounds of paper, 197 pounds of meat, and used an energy equivalent to 5.5 metric tons of oil. By close to 2008, those numbers increased to 635 pounds of paper, 275 pounds of meat, and an energy equivalent to 7.8 metric tons of oil.[85]

The United States is also perhaps the strongest proponent of contemporary globalization processes, which promote free-trade deals, increased consumption, materialism, and other pro-business and banking policies that are often at the expense of national sovereignty, public goods, and income equality. Commodification and consumption factor into core US-led globalization polices. US foreign policy is front and center of globalization as the push toward free markets and neoliberalization can include US military force or other means of coercion for countries that reject this path.

It also creates one of the largest carbon footprints. The US military creates more planet-warming GHG emissions than many industrialized countries, such as Portugal and Sweden. About 70 percent of the Pentagon's energy consumption comes from using and moving troops and weapons through the burning of jet and diesel fuel.[86] The militarization of US domestic and foreign policies need to be completely reassessed, not only for the influence of the military-industrial complex, the health of military members, the stability of countries and regions around the world, and the influence of foreign states, but also to address global warming. The battle to address global warming also demands a stop to immoral wars of aggression and a downsizing of the US military.

If the level of consumerism exhibited by the Western countries and the global wealthy was natural, then there would be no need to advertise and market. They actively condition consumption habits. The irony is that the US market is so important for international capitalism that if US consumption slows down, reverberations will threaten the global capitalist system. There are several threats to this system: unprecedented consumer debt, rising energy costs, the outsourcing of jobs, stagnant wages, and soaring rents. All of these challenge the economic feasibility of global capitalism. Because of this challenge, efforts by the United States to delay global action and international mitigation have proved successful for over a decade.[87] In many ways the United States has been regarded internationally, particularly in the Swedish press, as the great and overriding problem in the mitigation of global warming.[88]

Rest of the World

A US, Canadian, or EU-style mass consumption economy for the rest of globe is impossible without quickly destroying the planet; a four-to-sevenfold increase in the amount of resources would be needed for the rest of the world to enjoy the level of consumption in the United States.[89] If India and China develop the same ecological footprint as the United States, the entire planet will be stripped of its resources within a few decades. There would need to be two planet Earths just to sustain their two economies.[90] With Western countries primarily responsible for the climate crisis, climate justice would suggest they transfer resources to the Global South, or the "developing" countries, which are often the most affected by the crisis.[91]

Importance of Class

Income and class are also closely correlated with indirect individual emissions, which again are the emissions embedded in the consumption of goods and services consumed by individuals.[92] This means that individuals with the highest incomes or net worth consume the most and produce the largest carbon footprints.[93] The Earth's future productive and regenerative capacity is becoming even more challenged as global consumption is increasing for new, relatively small and wealthy groups (benefiting from neoliberal globalization) who consume at elevated levels. The wealthiest 20 percent of the world's population spends about 75–85 percent of total private spending, which translates to roughly the same amount of consumption.[94] According to Kevin Anderson, chair of energy and climate change at the Tyndall Centre for Climate Change Research, about "70 percent of global emissions of carbon monoxide come from about 20 percent of the world's population."[95] About half of all emissions come from about 10 percent of the world's population, while the poorest 3.5 million only account for a tenth, even though the poorest are the most threatened by global warming.[96] The discrepancies within China and India compared to the West are also noteworthy. The richest 10 percent of Indians have per capita emissions just one-quarter of the poorest 50 percent of those from the United States, while the average emissions of the richest 10 percent of Chinese citizens are about the same as the carbon footprint of someone in the poorest 40 percent of Europeans. Women in particular often bear the heaviest burden of climate change as they are generally more dependent on climate-sensitive livelihoods like rain-fed agriculture and water collection. The poor are also the least responsible. Someone in the richest 1 percent of the world's population uses 175 times more carbon than someone from the bottom 10 percent. According to one estimate, one billion people overconsume, while another four billion do not consume enough.[97]

According to Lucas Chancel and Thomas Piketty at the Paris School of Economics, with within-country inequality now making up 50 percent of global dispersion of carbon emissions, it is now "crucial to focus on high individual emitters rather than high emitting countries."[98] With a disappearing middle class, economic growth is now dependent on the consumption patterns a smaller group of wealthy individuals that are more likely to harm the environment. "And it is the most affluent who are . . . most able by virtue of their wealth to avoid the many inconveniences and life disruptions arising from the ecological crisis, and who are accordingly least inclined to be concerned about its material consequences" according to economics professor Eric Schutz.[99] He continues, noting that

> contrary to the drift of most mainstream discussion on this matter, the problem of social class is a fundamental element of both the global ecological crisis of these times and of the modern market society's difficulty dealing with it: the crisis will not be dealt with satisfactorily unless that problem is engaged as well.[100]

With wealth and class directly connected to the climate crisis, mitigating global warming also means fighting economic inequality. The wealthy consume the most and contribute significantly more GHG emissions compared to the poor and middle class. The elites overconsume the Earth's resources, while the poor, who often work to create the wealth that mostly benefits the elites, are only allocated a small percentage of the wealth.

Therefore, the planet will need to address existing class differences, but also existing policies that enable and continue to produce very rich people. The more rich people that society creates, the more they will consume. This is particularly troubling as the neoliberal era has produced a small number of very rich people, but mostly at the expense of everyone else's livelihoods. So, the rich are increasing. To address this crisis, political systems that do not allow public control over the distribution of wealth will need to be democratized to allow for this. Clearly, this will not be easy, but overconsumption by these groups will need to come to a halt. Perhaps it's time to take the lyrics to singer-songwriter Jeb Loy Nichols's song "To Be Rich Should Be a Crime" seriously in the fight against global warming.[101]

Unfortunately, all of the projected energy production and consumption data point to increases, not decreases, during the next few decades. A decrease is needed to avoid the point-of-no-return regarding the negative effects of global warming. The US Energy Information Administration projects increases in natural gas, crude oil, and lease condensate production for the next 24 years, when their prediction stops.[102] They also project a 48 percent increase in world energy consumption by 2040, which consists of liquid fuels (petroleum and other fuels), natural gas, coal, renewables, and nuclear energy. Nuclear production stays flat, while coal decreases, and renewables increase, at about the same rate as crude oil and lease condensate production. By 2040, natural gas and oil production will continue at the top.[103] The main problem here is that renewables still are at the bottom of production and consumption projections, even though they need to be at the top. Big oil's disastrous commitment to oil and gas means their low carbon investments are minimal, consisting of only 3 percent of their 2019 combined capital investment of $115 billion.[104]

Possible Environmental Purpose for Advertising?

The climate crisis has many causes that need to addressed. Besides the resistance of the fossil fuel industry, class differences, within-country inequalities, imbalances in global consumption, and systematic disparities created by capitalism all contribute to the crisis. In this chapter I argue that the relationship between production, consumption, advertising, growth economics, and global warming suggests production requires consumption, and ever-increasing consumption requires advertising. These processes are propelled forward by growth-economics, whose unchecked growth endangers the Earth's sustainability. Advertising has been a powerful tool for capitalist/growth-oriented relations that privilege economic gain over environmental conditions: the most familiar forms of advertising create demand in order to enable this environmentally harmful system. Sustaining increased production and economic growth would not be possible without current forms of advertising, and it is unlikely the climate crisis would have reached the current point without them. Advertising appears to have enormous power: our major challenge will be to figure out how to use that power to help save the planet, instead of hastening its demise.

In this chapter I argue that the four main human-made mechanisms that drive the climate crisis—production, consumption, advertising, and growth economics (with the first three mechanisms operating within the framework of the latter)—are *interdependent* to the extent that if any one of these is significantly altered, their collective output will most likely change. This suggests that significantly modifying any of the four mechanisms would likely alter the present trajectory of the climate crisis. Figure 1 visually represents this relationship.

To give an example, if suddenly all production processes are harmonious with nature, then the manner in which advertising and consumption occur would likely change. If people decide to only consume sustainable products, then production and advertising processes would likely change.

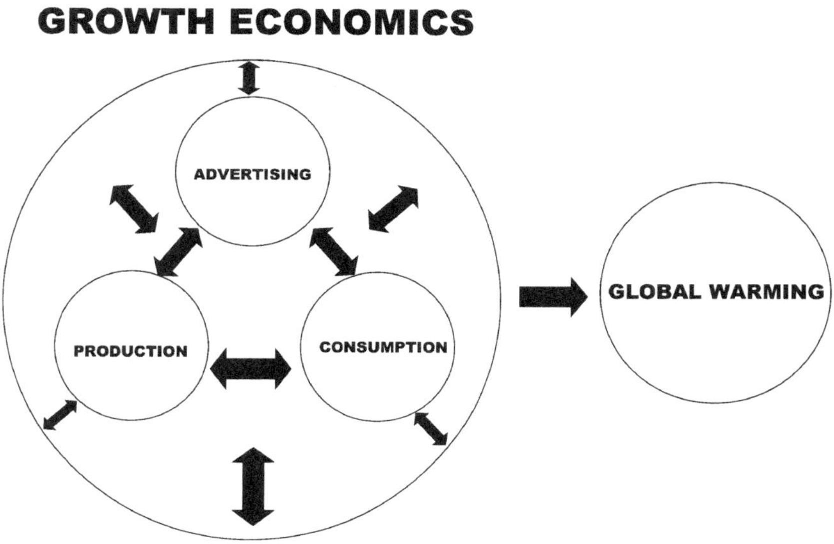

Fig. 1. Interdependent Mechanisms of Global Warming

If advertising policies are revised to only allow sustainable products to be advertised, then production and consumption will likely change. If citizens vote to eliminate their growth-oriented economic system, then production, consumption, and advertising processes will also likely change. In sum, substantially altering any of the four human-made mechanisms impacts the others, and thus their collective output.

Acknowledging the interdependence of human-made mechanisms concerning their influence in global warming has significance through their historical relationship with GHG emissions, as well as through their ability to create new strategies for future action. Advertising's agency in contributing to the global climate crisis is not to be ignored or underestimated. It is however, undertheorized, which is surprising given its impact over human influenced environmental degradation. The tendency to focus on advertising content or consumers instead of on the transactional and distributive processes within advertising undermines its potential agency.

The second argument presented in this chapter, which is premised on the first, is that there is a need for scholars and environmental activists to reassess the role of power relations and agency within advertising. While most critical scholars ascribe advertising solely with capitalism and negative agency, its interdependence with production, consumption, and growth economics suggests it has substantial aptitude to alter these mechanisms, and thus positively impact global warming. Revising advertising transactions and their distribution beyond pure financial exchange interpretations to new exchanges also based on environmental well-being may open new pathways for creating strategies to address the climate crisis. Scholars, politicians, and civic and activist groups may want to ask how changing advertising systems can assist in the transition to a postconsumption/post-economic-growth-system more focused on environmental sustainability.[105]

The logical trajectory of this chapter suggests that with some imagination and new policy recommendations, advertising could have an environmental and social purpose, and thus fulfill the view that communication practices be praxis-oriented to influence a more sustainable world.[106] Revising advertising to address the environmental sustainability of products and services to be advertised may help incentivize manufacturers to alter their production processes as desired by the IPCC.[107] A new system that encourages environmental sustainability instead of overconsumption could contribute to future policy, or at least assist in the transition to a more environmentally sustainable economic system. The next chapter extends this discussion by rethinking advertising to introduce the beginning of a new advertising model that can help mitigate the climate crisis.

CHAPTER FIVE

Conceptualizing a Mitigative Model of Advertising

Influenced by the argument that communication practice must be praxis-oriented in order to influence a better and sustainable world,[1] this chapter is about the need for praxis, action, new thinking, and policy solutions given the gravity of the crisis and the need for immediate action. I rethink advertising to propose a mitigative model that may begin to fill the much-needed gap in studies that directly address incentives for manufacturers of products and services to better control the GHG emissions generated by their production processes. Developing a new advertising approach that considers environmental sustainability would help mitigation efforts and effectively complement the strategy of curtailing demand, as suggested by the IPCC.

The remaining sections first discuss governmental regulatory bodies, along with some of their historical precedents, that can help address the crisis. Next I examine the idea that green advertising and self-regulation can effectively address the crisis and present an argument connecting advertising with other industries that cause environmental and human damages. Many of these industries are regulated with taxes and fees, while advertising is not. After this, I present a mitigative advertising model in six components and explain how it could be operationalized. There is much support for and resistance against new mitigative initiatives, which suggests change won't be easy. As ridiculous as this may sound to many, advertising could potentially play a part in saving the planet, instead of expediting its demise.

As the previous chapter noted, advertising acts as an agent to create demand by trying to persuade citizens to consume more and more energy-using goods and services; and in growth economies, economic performance is directly connected to levels of consumption.[2] According to Graham Murdock, professor of culture and economy at Loughborough University, the evidence linking the escalation of climate change to the expansion of capitalism is now incontrovertible.[3] In this sense, commercial service/commodity advertising plays an important role in terms of contributing to the climate crisis.[4] A perpetual growth system dependent on consumption ultimately stretches the natural environment to its limits, thus harming the Earth's bio-capacity and its occupants.[5]

Advertising continues to expand around the world while the consumption of goods and services is also increasing. At the global level, the production and consumption of these goods and services has dramatically increased (on a per capita basis) over the past several decades.[6] Again, this is extremely problematic because they are central to the production of GHG emissions, which significantly contribute to the global climate crisis.[7]

To address overall global warming, the IPCC sees advertising, and the production and consumption of goods, as targets in need of reform. It recommends mitigative strategies that decrease GHG emissions by reducing product demand, extending product life, and limiting unending economic growth.[8] For the IPCC, the challenge is that there are currently no clear incentives to improve suppliers' manufacturing processes or reduce product demand.[9] Regarding this as a significant policy knowledge gap, they call for new policy recommendations to address the needs outlined by the IPCC.

Degrowth strategists recommend limiting or eliminating marketing and advertising altogether in their approach to the crisis.[10] They suggest establishing very restrictive criteria for allowing advertising in public spaces. They also recommend creating committees to control the quantity and quality of advertising in addition to taxing advertisements according to the committee's objectives.[11]

However, both the IPCC and degrowth strategists offer limited analyses of advertising, as they attribute it with only negative agency. Most critical scholars of advertising also envision advertising with negative agency.[12] They all tend to downplay or ignore the transactional and distributive aspects of advertising. Downplaying the transactional and distributive aspects of advertising is unfortunate, as they can have a strong impact on the human and environmental consequences of products and services that are advertised through media. This book emphasizes these aspects of advertising as opposed to actual advertisements themselves. The content of advertisements is mostly irrelevant to this book. Rethinking the transactional and distributive aspects of advertising is the key to using advertising as a mitigative model.

Government Regulation Efforts

Historically, consumer legislation, law and regulations enjoyed some success in the early part of the twentieth century with various food, drug, and trade acts,[13] and again finding some traction in the 1960s, which included laws to protect children, consumer credit, the labeling of goods, and other safely initiatives. The US government has numerous government agencies whose missions are to regulate the media, as well as to protect consumers and the environment. For example, there is the Federal Trade Commission (FTC), the Federal Communications Commission (FCC), the Food and Drug Administration (FDA), and the Environmental Protection Agency (EPA), just to name four. It is important to note the United States has a history of protecting various aspects of daily life and that there are several governmental bodies that could regulate advertising vis-à-vis the global climate crisis.

However, while government regulation of advertising never addressed harmful production processes, it was bold enough to stop numerous companies from advertising when they broke other regulatory policies. For example, the FTC often issued "cease and desist" orders to companies who made deceptive claims (via the concept of unfairness) in their advertising. In fact, there were some 614 orders delivered between 1963 to 1969.[14] Even though these orders were not the most effective as by the time the companies received them their campaigns were already over, they at least existed to benefit consumers. By 1980, the Reagan administration limited the FTC's ability to regulate advertising.[15] Revisiting the "cease and desist" orders in a more effective way could help prevent companies with harmful production processes from advertising their products and services.

The UK's Advertising Standards Authority is also regulating, by eliminating, advertisements that are regarded as harmful for society. Their focus has been on sexist advertisements, instead of on advertisements for products or services that cause environmental harm. This means that British companies will no longer be able to create gender-stereotypical advertisements in addition to advertisements that suggest transforming your body will make you romantically successful. The justification for changing these advertisements is that many fear such depictions contribute to pay inequality and cause psychological harm, among other negative effects.[16]

Other countries and cities have banned advertising outright. In 2006 Sao Paulo, Brazil passed a law banning advertising from its streets, while Grenoble, France became the first European city in 2014 to ban advertising and replace billboards with trees. Their goal was to develop areas for public expression away from commercial representation.[17] While this is a much needed and excellent start, there is still a major need to tie advertising in with additional, or more mitigative, approaches to address global warming.

Failure of Green Advertising and Self-Regulation

Advertising industries have successfully lobbied for limited and self-regulation policies of their business. This is a significant impediment in the battle to address the climate crisis. However, many view "green advertising" as the solution to the challenge of advertising, consumption, and the climate emergency. Green advertising researchers often examine advertising channel selection, claims, message strategy, attitudes, and purchase intentions, while largely viewing advertising as a tool to achieve corporate goals of creating a positive image with brand equity.[18] According to scholar William Kilbourne, this research "tends to be fragmentary and highly specific with common goals focusing on the identification of ecologically conscious consumers as a target market or the development of scales to assess the level of environmental concern among consumers."[19] Green advertising tends to focus on advertising content and consumers, rather than on the distribution and transactions that also make up advertising. Unfortunately, green advertising, like self-regulation and the lack of significant governmental oversight in advertising industries, has made little headway in addressing the environmental challenges facing the planet. Marketers often tend to sideline overall ecological concerns,[20] and new approaches and thinking are still needed. Sadly, mass communication scholarship is sparse on policies or regulations that address the production processes of items or services to be advertised.

The inability of self-regulation within advertising to transition to a sustainable and harmonious relationship with nature highlights the power that industries have over regulatory schema. The FTC currently has a "Guide for the Use of Environmental Marketing Claims," otherwise known as the Green Guides, which sets guidelines for corporations to not deceive consumers about environmental claims.[21] However, the guide only examines claims and not actual production processes with regard to sustainability. The world's foremost business organization, the International Chamber of Commerce, has also created a "framework for responsible environmental communications," but this framework is also self-regulatory and only examines advertising claims and not production processes that can harm the planet. Their framework is designed to create a "sound" code for environmental advertising claims and practices. It is also predicated "on the principle that freedom of commercial speech in the sale of all legal products and services is a fundamental tenet of free markets, and that free markets promote innovation and competition, benefiting consumers by giving them more choices and efficient pricing."[22]

Indeed, after several years of Green Guides, International Chamber of Commerce frameworks, and more than a decade of green consumerism, each has failed to produce significant inroads into the unsustainable nature of consumption and production.[23] Many scholars see green advertising as having a negligible effect on mitigating the climate crisis.[24] Moreover, when promoted as a sort of solution to environmental problems, green consumerism may be more likely to block real solutions as they do not address the cause of the degradation. This is partly because green advertising and consumerism exist within a neoliberal, capitalist context that prioritizes profit over saving the environment.[25] Moreover, green advertising outsources responsibility onto consumers for a problem that can't be fixed by individual behavior.[26] Scholars Richard Maxwell and Toby Miller note that "the discourse of consumer ethics can be a poorly disguised effort to slough off responsibility for saving the environment onto individual consumers who, as we have argued, are in no position to effect change on their own at the scale that is needed."[27] Indeed, larger structural challenges facing consumerism and the environment are unlikely to be fixed by individual choices.[28] However, the outsourcing of consumer responsibilities benefits industries because it deflects attention away from the corporations that cause many of the environmental problems, as well as avoids discussing the role of government regulation.[29] This is unfortunate because government regulation of corporate activity is more likely to be effective in addressing the climate crisis.[30]

Regulating Advertising

In this section I argue there is a need to rethink advertising regulation by expanding it to the transactional and distributive aspects of advertising. It needs to address the way advertisements are bought, sold, and distributed.

These aspects are regulated in a number of other industries because they can harm people and the environment. This means that not everyone can sell or purchase certain goods. For example, to acquire or purchase alcohol, tobacco, prescription medicine, explosives, and in some states ammunition or weapons often requires identification, with age limits, a background check, and perhaps some sort of testing or licensing on behalf of the seller. The receiver and the seller need to meet certain requirements, set up by society through laws and policies, in order to partake in the transaction.

Similarly, in order to distribute items that are harmful, or potentially harmful to people and the environment, society has also created a number of laws and policies to help reduce potential damage. For example, not everyone can freely distribute the large fireworks, prescription medicines, certain kinds of gasses, weapons, explosives, and chemicals mentioned above. Even restaurants can't serve any kind of food under any condition. Distribution for all of these areas is regulated because the items are, or can be, dangerous under certain circumstances and would need to pass some sort of safety process in order to be dispensed.

The regulation of the transactional and distributive aspects of these industries do not apply to advertising, even though the products and services being advertised can also negatively contribute to environmental and human maltreatment. Instead, only the visual aspect of the advertisement, through the content, is somewhat regulated. For example, advertising nudity appears to be regulated in certain areas. Viewing advertisements with naked people in public areas is generally difficult to find, even though one could make an argument that it is hard to tell as a lot of them are very close to naked, as male and female models in multiple advertisements are *almost* entirely naked. Nonetheless, this aspect of advertising appears to be somewhat regulated. Obviously, not all aspects of advertising are dangerous or harmful all of the time, just like not all aspects of food distribution are dangerous all of the time. However, advertising transactions that don't address harmful production processes are detrimental for the environment. If better regulated, they may be able to contribute to mitigating the climate crisis.

Role of Fees

One way of addressing harmful production processes is by creating fees for products and services that are not environmentally sustainable. Creating fees and regulating various aspects of society due to potential harm is nothing new. For example, cigarettes and alcohol have been regulated and taxed for decades because they can be harmful for humans. Taxing advertising is also nothing new. Although internet advertising is often excluded to the detriment of revenue for states, taxes on advertising exist in Austria, Italy, Greece, Belgium, Estonia, Croatia, Sweden, Portugal, and Romania.[31] Many already consider advertising a private tax on citizens, where consumers pay more for their products in order for the companies to pay to advertise their goods and services, which adds up to $700 billion a year.[32] Indeed, taxing can be an effective way to help mitigate various aspects of the climate crisis. An interesting study found that including the social health cost of red and processed meat consumption into their prices could lead to significant environmental and health benefits.[33] In essence, this meat tax would likely help save lives, positively impact the environment, and cut health care bills.

A recent poll indicates that the US public is also supportive of taxes that address the climate crisis. According to an article in the *Guardian* newspaper that examined studies and polls mostly coming out of Yale University, American citizens are indeed willing to pay a carbon tax of $15–20 per month on their energy bills to combat global warming. Moreover, 78 percent of registered American voters support taxing or regulating carbon pollution, or both, including 67 percent of Republicans and 60 percent of conservative Republicans.[34] Moreover, 6 in 10 Trump voters support taxing or regulating the pollution that causes global warming.[35] However, who is exactly supposed to be taxed and who will ultimately pay would likely go through legislation and perhaps the courts as industries, such as the fossil fuel industry, would prefer not to pay for their contribution to the damage. That may leave citizens picking up the tab, like they did in South Florida. Miami voted to tax itself $200 million to mitigate sea level rise,[36] instead of directly taxing the industries that were responsible for the majority of global warming and thus of sea level rise.

Countries, states, and regions can change or develop government agencies relatively quickly to address societal needs. For example, in the United States, massive efforts have been under way in several states to regulate and tax legal marijuana. In California alone, the Bureau of Cannabis Control has 31 testing labs across the state.[37] The massive amount of regulation, testing, and enforcement needed to manage the legalization of marijuana is likely to be much more complex than the regulation of advertising based on production processes.

The point of this example is that governments move and do things for the public. With public and corporate awareness increasing, it is time to address global warming with new forms of regulation. The current system perpetuates the climate crisis. Directly taxing manufacturers of products and services that harm the environment is unlikely to happen given the power that these industries have over governments. The strategy of having the US government tax harmful industry production processes to help mitigate the climate emergency has so far failed under these conditions. While these taxes definitely need to exist, using the advertising system as an alternative or a complement constitutes a powerful and effective strategy in itself. A mitigative advertising model can help and could be run through an existing or newly created government agency to oversee this process. Perhaps the FTC would make the most sense as it is already responsible for regulating advertising; however, in its current form, it is less interested in dramatically altering the system and is more attuned to perpetuating the present advertising reality.

Introducing a Mitigative Model for Advertising

What would an advertising system look like that is more harmonious with the environment? What would a system look like that could aid in the transfer to a post-economic-growth-focused and consumption-based economic system? There is a need for a new system that does not harm the Earth by encouraging overconsumption. Instead, the system needs to encourage sustainable consumption and production, which would give advertising a social and environmental purpose. The logic would follow that if many manufacturing businesses' production processes encourage the depletion of the Earth's resources and create unsustainable GHG emissions, then an advertising system needs to be created that can target these harmful production processes by incentivizing companies to change.

When economists Paul Baran and Paul Sweezy wrote that society creates the foundation by which advertising exists, they partially meant the ultimate authority in creating an advertising system lies with the public.[38] In effect, the public has a democratic right to set the parameters of how advertising should function for society. When scholars Justin Lewis and Tammy Boyce pointed out that the "media and telecommunication industry fueled by advertising and profit maximization" is "part of the problem, rather than part of the solution,"[39] they suggested that large-capital control over the media has determined the use-value of the advertising system. It is their system, they own it, and it benefits them over the health of the planet. Therefore, the role of advertising, as the main mechanism by which a production-driven consumer/growth-oriented economy creates demand, must be reassessed given the antagonism between the depletion of the Earth's resources and the sustainability of the planet. Because many outcomes of global warming can still be avoided, reduced, or delayed by mitigation,[40] it is time to rethink the advertising system. As Lewis and Boyce noted, "just as we caused it, we also have the capacity to control it."[41] However, we cannot solve a problem with the same kind of thinking that created it in the first place.

In response to the IPCC's need for new incentives,[42] a mitigative advertising model would create incentives for companies with unsound production processes by increasing rates on their advertising costs.[43] Altering the advertising system with additional fees would establish a motivation for manufacturers to use nonpolluting production processes and sustainable material. If effective, it would help cultivate products and services that maintain a healthy and sustainable planetary ecosystem.

A mitigative advertising model focuses on the transactions enabling the distribution of material commodity and service advertising. A material commodity consists of any nonconsumable item (i.e., not food or drink) that is manufactured with components taken from the Earth and that is sold or traded on the marketplace. Services are intangible products, such as cleaning, transportation, and banking, where no transfer of possession or ownership takes place when services are sold. With US manufacturing and agriculture industries declining, services now represent more than 80 percent of GDP, with information services signifying the largest share of value-added GDP along with 59 percent of the US workforce by 2000.[44] Much of this likely constitutes banking and investment services where money is transferred between entities. The production and consumption of products and services still drives most GHG emissions. While there are myriad varieties of advertising, this model focuses on material commodity advertising and service advertising, and does not focus on political, food, sports, or other kinds of advertising. Additional scholars may want to reexamine these areas, given that they each interact with global warming differently. I focus on commodity and service advertising because they constitute the largest percentage of advertising content in consumer-oriented economies,[45] while their production processes significantly stress the Earth's resources.[46]

It is important to highlight this aspect again. Regulating the actual production processes of services and commodities would be ideal; however, if the political and economic forces opposed to this continue to prevail, then regulating the advertising of goods and services could be an easier and useful alternative, or a complement to production-process regulation. For the mass communication field, advertising not only contributes the most to global warming, it also appears to have the most potential to help mitigate the crisis.

Two-Phased Strategy

This section outlines a broad, two-phased strategy to help mitigate global warming.[47] It first establishes criteria to assess the production process of products to be advertised in the marketplace. Second, the model creates a new rating system for advertising, based on the amount of environmental damage created by the production process of the commodity or service to be advertised. The model posits that higher rates should be levied on businesses that infringe any combination of the six components described in this model, while companies with sustainable production processes would pay lower or no additional rates.[48]

Manufacturer Assessment Process

Six components begin a discussion regarding what production processes would need to be assessed in order for an advertising system to measure and address unsustainable manufacturing production processes. Once assessed, each manufacturer wanting to advertise could be given a score based on the six components. The score would determine the rates the companies would pay in order to advertise through advertising agencies and the media. The assessment results would then be used by advertising agencies and media organizations, as they are the intermediaries that connect the manufacturer to the larger public through various media forms. A mitigative advertising model would use the advertising distribution system as a gatekeeper to issue additional rates in order to incentivize manufacturers to produce environmentally sustainable products that are less likely to harm the Earth's ecosystems.

Model Components

First: Measure the self-sufficiency of the production process and apply rates accordingly. This means measuring the degree to which the commodity to be advertised is made with renewable energy and resources. Here the incentive is to create pressure for the production process to be as self-sufficient as possible. In this case, the self-sufficiency of the production process of the physical item for sale, or for the service, and also for the workers who create the products, are all important. This means examining renewable resources, construction, energy production, food generation, housing, water, and so forth. Production process emissions that do not exceed the renewable assimilative capacity of the local environment are healthier for ecosystems.[49] Creating a product or service produced from renewable resources at rates lower than the resource's replacement rate is also more ecologically sound.[50]

This component also acts as an incentive for businesses to promote and invest in the creation of solar, wind, tidal, or other forms of renewable energy. Different rates could be imposed for those commodities or services that exceed regeneration rates of materials used in the production process.

Second: Measure the percentage of recycled and recyclable materials that constitute the commodity to be advertised. Also, measure the product's ability to be recycled. The goal here would be to encourage the use of recycling. Recycling of human waste into crop production can also assist agricultural production processes. Renewable resources that preserve local resource bases and that do not endanger other species or plant life or the general environment will need to be rewarded. A scale system of new advertising rates could be levied on the manufacturer with higher rates charged to those businesses that do not recycle or use recyclable materials in their production processes.

Third: Measure the percentage of workers/employers/employees who use public transportation or collective transportation, or walk or bicycle to get to work. This not only lowers carbon emissions but can also be used as an incentive for businesses to support the creation of more public transportation. Here, advertising rates could be applied to the manufacturer of the product to be advertised based on the mode of transportation the employees use during their workdays.

Fourth: Create fees for businesses wanting to advertise that have not divested in holdings from fossil fuel companies. Reports from Oil Change International and the IPCC note it's best to keep oil, gas, and coal in the ground, while all new fossil fuel infrastructure and industry expansion should stop.[51] Oil Change International notes that the potential carbon emissions from the fossil fuel industry throughout the world would take the planet beyond the 2°C warming limit, while the current reserves would take the world beyond the 1.5°C limit. Businesses in the fossil fuel industry should also incur higher advertising rates, given their elevated role in contributing to climate change. This incentive puts pressure on fossil fuel companies to stop what they are doing and perhaps go into renewable energy instead. If these companies moved into renewable energy, there may be a need to break them up through antitrust measures, or for states to own all or part of them to cultivate competition or accessible pricing, or both, for the public.

Fifth: Measure the degree to which the production process of the commodity or service to be advertised is nontoxic for the environment. This would require understanding the amount of carbon and greenhouse gas emissions generated during the production process. The degree to which the good or service to be advertised is biodegradable is also important. Natural materials that can compost or break down require less energy than recycling. A system of rates could be levied on the manufacturer, with higher rates for those businesses responsible for elevated emissions, along with those that create nonbiodegradable products.

Sixth: Measure the degree to which the production process of the commodity to be advertised is intentionally created to break down or expire over a short period of time, and thus require replacement. This is called planned obsolescence. It is often more profitable to design an item with cheaper components to break down after a few years, instead of building a product to last for several decades.[52] Creating a product to expire or become useless after a short period of time also promotes waste. For example, Apple does this frequently. It spent almost $1 billion on advertising its goods in 2012, but what they were advertising were basically newer and improved versions of goods most people already had.[53] Indeed, hundreds of millions of still working items, constituting billions of pounds of hazardous waste, are thrown away every year.[54] By 2007, between 20 and 50 million tons of e-waste were generated annually, with the majority coming from televisions, computers, and cellphones.[55] The planned-obsolescence business strategy creates unneeded waste at the expense of the environment. Therefore, additional rates could be levied on those manufacturers that intentionally create products to expire, become useless, or break down within a short time frame.

In order for this model to function, a neutral third-party entity would need to assess companies before they could advertise, and public universities may be the most reasonable organizations for this need. Perhaps initially based on market share, those companies with the largest segments wishing to advertise would need to first pay a fee to have their production processes assessed. This assessment would then determine any additional advertising costs for the companies. If universities conducted these assessments, the revenue generated (from the assessments, not the additional advertising fees) could offset education budget cuts, while helping fund new strands of ecological research, and thus contribute to the public good.

Corporate and Public Resistance and Support

There appears to be both resistance against and support for sustainable initiatives. Advertisers and media tend to be hostile to criticisms of consumerist ideologies or arrangements where advertisers significantly benefit.[56] In fact, "no other group in democratic societies commands the same power as marketers to speak," according to professor Jonathan Hardy of the University of East London.[57] Clearly, suggesting regulation based on the extent to which production processes influence environmental harm would likely garner a negative response from some industries. This would be nothing new. When critics wrote how advertisers were manipulating people into buying products by using persuasive strategies in the 1960s, advertisers responded by denouncing their work as propaganda.

According to scholar Ernesto Laclau, "That is the reason why, when new social forces try to bring about fundamental social changes, they necessarily clash with the existing institutional arrangements. Corporate power will defend *à outrance*, a purely institutional outlook, which represents the existing status quo."[58] For example, car companies, as one of the largest advertisers in the United States, would likely push back against a new advertising system that took production processes and fossil fuels into account. They have played a major role since the 1920s by rallying against sustainability initiatives. They purchased several public trolley systems only to dismantle them; they lobbied for increased road support, and challenged organizations that fought against car usage. To promote individualized auto-centric societies, they spent $15.6 billion on ads, $67 million on lobbying, and $19 million on campaign contributions in the United States alone during 2008.[59]

However, companies that continue to resist mitigating global warming by not changing their production processes risk loss of competitive positioning and negative reaction from buyers, pressure groups, and some media and politicians.[60] Many companies see this, as not all companies are opposed to sustainable modes of production and many are already changing the way they create products and services. For example, a study from the global climate change disclosure nonprofit CDP notes that more than 1,200 corporations either have an internal carbon pricing plan or are about to put one in place.[61] Many companies are beginning to realize that they can gain competitive advantage by incorporating more environmentally friendly products into their production lines.[62]

Advertising industry professionals and consumers also are also concerned about the crisis. One study found that about 75 percent of people working in the advertising industry somewhat or strongly agree that advertising leaders are concerned about climate change.[63] Another study found that 89 percent of consumers think that companies should try to solve social and environmental problems. Moreover, 67 percent indicated that they trust sustainable companies more than those that exhibit nonsustainable practices. According to the director of the Havas Media Group, sustainability is a platform that can "construct relevant brands, because it makes them more human and reconnects them with people's preoccupations and aspirations, infusing the brand with meaning and purpose, an ideal and shared values."[64]

Perhaps as a result, more and more companies like Google, Apple, and IBM are now promoting various aspects of responsible production and consumption.[65] Google and Apple are working to exploit pig manure as a potential energy source.[66] There may be enough pig manure in the United States to create sufficient methane gas to address the massive energy-consumption needs of data centers.[67] Apple also asked Ecoalf, a sustainable company that designs urban wear and other products out of recycled waste, to make their Mac Air and Mac Pro computer cases.[68] Another example has IBM moving to Canada in order to take advantage of the cooler climate, which translates into cooling their data centers for 210 days a year without running energy-consuming chillers.[69] Google also said it would try to be completely powered by renewable energy by 2017.[70]

These initiatives sound great. They begin to address the need for sustainable industries. However, many are still connected to other industries for energy or their products require a lot of energy, or both. This is troubling since there may not be much renewable electricity left for others if large quantities of renewable energy go to the businesses, as well as to the products they sell. In 2015, Google purchased 5.7 terawatt hours (TWh) of renewable electricity, but that was slightly less than the total amount of energy generated from the UK's solar panels that year, which was 7.6 TWh. That only left 1.9 TWh for everyone else.[71] That meant that the amount of energy they are buying, versus the displacement of renewable energy that could be going to other needs, like heating or household electricity, is becoming an issue. However, Google also views nuclear energy as acceptable and plans to potentially use it in the future, even at a time when Fukushima and other risks of nuclear energy are becoming more and more concerning for people around the world.

Need for New Policies

While communications policy researchers can often exhibit self-castigation given the challenges they face, some suggest there is a significant need to reinvigorate this much-needed voice.[72] With public and corporate awareness increasing, perhaps addressing these contemporary challenges by introducing new models and policies could gain momentum. If governments resist the concept of an advertising system that addresses product production processes (which is the most likely scenario for the federal government in the United States), then civic groups may want to think about passing local laws in support of such a system, or pressure political parties to adopt the model within their platforms.[73] According to a report by the United Nations Environment Programme, "action by non-state and subnational actors (NSAs), including regional and local governments and businesses, is key to realizing greenhouse gas emissions."[74] Indeed, climate change is a collective-action problem and therefore requires the public to facilitate policy change.[75] This means local municipalities, parishes, and counties could decide to regulate the manner in which local print and broadcast advertising is distributed.

The digital world may require a different strategy because it's more difficult to regulate given its scope. Digital advertising can be local, national, and international, unlike the local newspaper and television station. Google and Facebook control over half of the global digital advertising market,[76] and together with Baidu, Microsoft, Yahoo, Verizon, and Twitter generated $132.8 billion of internet ad revenue in 2016, or nearly three-quarters of all internet ad spending, and nearly one-fourth of all ad spending.[77] These companies will need to work together with public, communications, and environmental groups to adopt a more eco-friendly digital advertising model if governments are ineffective. If these companies are not interesting in self-regulating,[78] organizing civil-society boycotts of their businesses may get their attention and interest. Indeed, boycotts can work. For example, Google recently made efforts to change its advertising policies in response to a growing boycott of the company's platforms from leading brands and advertisers, which included Marks & Spencer, the British government, and McDonald's.[79]

Changing advertising in favor of a healthy ecosystem may be easier than anticipated largely due to the oligopolistic nature of the advertising industry and most other industries that manufacture material items. For example, at the turn of the millennium there were just five advertising firms doing more than three-quarters of the world's business, with eight companies receiving 75 percent of television advertising in the United States.[80] By 2013, four major advertising groups dominated the world: WPP Group (British), Interpublic Group (US), Omnicon Group (US), and Publicis (France).[81] The media also largely exist in oligopolistic markets where a small number of major players dominate. This means that advertiser-sponsored media, product placement, and other advertiser-related phenomena not considered "direct" advertising can also be addressed with the mitigative advertising model. The small number of media corporations makes this less complicated for now. However, this does not suggest that citizens should prefer a consolidated media industry. Most would correctly prefer to break up this oligopoly.

Most other industries, like vehicle, computer, phone, other electronics, and clothing manufacturers, also exist in oligopolistic markets. According to one study, roughly 1,318 companies own about 80 percent of transnational corporate global revenues.[82] This concentration means that the players are powerful, but the number of targets is limited, and they are somewhat vulnerable to competition with one another. If mitigative advertising practices can be shown to give them an edge, they may be likely to accept the changes. The resources needed to implement a mitigative advertising system would likely be reasonable given these levels of concentration.

Both the proposed offline and online advertising strategies would require a well-planned grassroots campaign using numerous strategies and tactics that incorporate various public relations and advertising approaches. Strategists may want to start with larger cities, where the majority of potential consumers live, while simultaneously beginning to communicate directly with Google and Facebook. If cities like San Francisco, Los Angeles, New York, Chicago, and Atlanta were to adopt mitigative advertising models, one would likely begin to see companies make serious headway toward changing their manufacturing processes.

Nonetheless, for any model to alter the present system, ecologists, environmentalists, and advertising scholars would need to collectively discuss policy recommendations. Creating and prioritizing specific variables, like the six components presented here, would need to be reassessed and updated. The environment is fluid, not static. For example, if the production process of the product to be advertised pollutes the environment more than when employees drive to work (instead of bicycling or walking), then that dimension would need a higher value and thus a higher advertising rate. The public could set the rates. For example, using the six dimensions to determine advertising rates could add up to a totality of up to 40 percent of the total cost to advertise under current conditions. For a million-dollar television ad that infringed all six dimensions, the advertiser would need to pay $400,000 above the million dollars for the ad. This sum would go to the government agency in charge of its oversight. With advertising costs conservatively running at $280 billion in the United States,[83] bringing in just 10 percent of this total would translate to $28 billion. This is a significant figure given that the World Bank is planning on using $100 billion in climate funding from 2021 to 2025 to cut emissions and to protect people from floods, storms, and droughts.[84]

It is likely the ideas presented in this model would not only incentivize manufacturers to create more eco-friendly services and products but the model would also bring in substantial revenue. Income from this model could be divided up in at least three ways. Some could be given back to the public, like the Canadian government is doing with its recent federal carbon tax initiative. In their case, all of the taxed money is distributed back to the provinces from which they were generated, where 90 percent of the revenues are returned to individual taxpayers. The rebates are anticipated to exceed the increased energy costs, which means that families receive a larger sum from the rebate than what they paid for the increase in costs due to the carbon tax. The net financial benefit to taxpayers is to grow over time.[85]

Besides being given back to the public, the money generated from the model could also be used for further mitigation and remediation, or it could also be used to improve democratic media infrastructures. Perhaps it could be used for all three possibilities. While I argue for advertising reform in this chapter, chapter 10 argues that addressing climate change also means supporting journalism and media reform. Establishing better publicly funded press and media systems is likely to improve climate change journalism, as well as help improve countries' standings on democracy indexes. Indeed, countries with "the freest press systems are also the nations that make the greatest public investment in journalism and therefore provide the basis for being strong democracies," according to professor Robert McChesney.[86]

Another benefit of having more public media is that they are less funded by advertising interests, which means private media funded by advertisers have a fixed stake in the promotion of a consumer-based economic system. The revenue generated from a mitigative advertising model would be a bonanza for the protection of the environment and for the democratization of the media. Taking a small slice of the $28 billion would go a long way for media reform given that the US Congress allocated only $445 million in 2014 for the operating budget to the Corporation for Public Broadcasting, which provided 15 percent of PBS's annual budget and only 2 percent of NPR's budget.[87] Chapter 10 describes additional ways of funding public media.

The mitigative model proposed in this chapter is the first to directly respond to the IPCC's stated need for new policies that address incentives to improve production processes, decrease GHG emissions, and reduce product demand.[88] It outlined a new approach that questions the present relationship between advertising and the environment. It is one of the first to go beyond the established focus by examining the distributive and transactional aspects of advertising and their relationship with the climate emergency. Last, the chapter's praxis-orientation suggests that the field of advertising can influence efforts that address the crisis by providing a practical and feasible contribution.

This chapter concludes the section on advertising, global warming, and reform. The next section of this book focuses on digital communication technologies and their ecological impact. Chapter 6 extends the discussion of advertising into the digital world by describing how the online infrastructure works for advertising interests through its surveillance mechanism, which unfortunately helps increase overall consumption, further fueling the climate emergency.

III

The Digital Era and Its Contribution to the Crisis

CHAPTER SIX

Digital Era Advertising, Surveillance, Exploitation, and Inequities

The future belongs to marketers and media firms.
—Joseph Turow[1]

While the previous chapter proposed a new model for advertising that would favor environmental sustainability in the face of the climate emergency, this chapter goes into more detail regarding the digital era's processes that enhance consumption and thus contribute to global warming.[2] I examine online business surveillance and individualized pricing tactics in the postindustrial, advertising-driven digital era. I ask, what will their effect be on consumer sovereignty, consumption trends, and general economic inequality? These are three important areas that influence global warming.

The chapter's first section examines scholarly debate on the potentially empowering and liberating aspects of digital technology. I look at individualization and aspects of the internet's relationship to power, class, and economic inequalities. Next, I describe the incredible growth in online advertising, overall consumption, and e-commerce revenue during the past few decades. These trends emphasize the overall success of the sales effort within the digital era. The remaining sections constitute the main elements of analysis. They examine how commercial surveillance, tactics, and technologies interact with online users, as well as how they challenge consumer sovereignty and agency.[3] Particular attention is paid to price customization, which creates unique online prices for different people who view the same product. Many online practices and developments exploit online user labor and consumer agency, as well as exacerbate economic inequalities. How these aspects of the digital era impact the climate emergency are then discussed in the next chapter.

In this chapter I argue that the postindustrial, advertiser-driven digital era exploits online users in multiple capacities, which reduces consumer sovereignty and agency. The result of this exploitation enhances consumption, significantly increases online profits, and continues to consolidate wealth in the hands of a few. Numerous mechanisms propel this process. The privately owned oligopolistic/monopolistic information and technology sector enables the surveillance and exploitation of internet users. This sector collects and sells user data to businesses and advertisers, which in turn use the data to individualize content in order to more effectively advertise, set prices, and sell goods and services. These developments aid overall consumption and maximize capital extraction by those who already possess the means of production, which in turn heightens their power and influence within society. As the individualized/customized online phenomenon moves beyond its initial stages of development and becomes more ubiquitous, the economic effects will likely broaden, comporting with capitalism's overall historic trend of exploitation and class division. Studying online surveillance, data collection, and advertising in this context is important, as information asymmetry has the potential to exacerbate overall social economic inequity.[4]

Theoretical Perspectives on Technologies, Inequities, and Individualization

Looking back in time, few new technology theorists predicted the exploitative potential of the internet. Instead, many believed it would empower users and improve consumer welfare. For example, Nicholas Negroponte saw online technology enabling individuals to filter out what they wanted online, in order to receive information that was most important and relevant to them.[5] Here, users became empowered and more in charge of their online experiences, thanks to new technology. Yochai Benkler, like Negroponte, also believed that individuals would hold power over their media destinies in the twenty-first century.[6] Andrew Shapiro went further, predicting that the internet would allow individuals to take power away from the media and other institutions,[7] while others thought it would shift control from supplier to consumer, perhaps by giving voice to the powerless.[8]

Utopian scholars also believed the internet would allow greater freedom, liberate people from oppressive social norms, and expand democracy by undercutting authoritarianism and social divisions.[9] New tech CEOs concurred, often believing the internet would solve problems all over the world while reducing poverty.[10] Moreover, even the US government believed big data had the ability to improve the lives of *all* Americans.[11] In general it was believed that access to information through information and communication technologies would improve the quality of individual lives around the world.[12]

A few decades later, after the internet began to morph into what it is today, many scholars began to find the statements from utopian theorists erroneous and out of context. Today, many democracies are in decline; authoritarianism, racism, and nationalism appear on the upswing; and internet corporations allow false and misinformation to spread freely, while they simultaneously censor other content they deem inappropriate. Suggesting that the macro effect of the internet and new technologies under global capitalism has created a more democratic and egalitarian world is clearly naïve and misleading. It's less likely to happen under current conditions. For example, Joseph Turow, professor of communication at the Annenberg School for Communication, believes the internet does not necessarily empower citizens and consumers. He notes that advertisements, information, and entertainment are increasingly customized to individuals, based on a vast amount of online social profile information collected without users' knowledge or consent. He argues that the burgeoning information asymmetry is unlikely to empower internet users.[13] Other scholars see individualization resulting in the creation of idea cocoons, or internet filter bubbles, where online users become fragmented and largely stuck in their own information worlds.[14]

As *Dollarocracy* authors John Nichols and Robert McChesney note, "Google search results for the same entry generate different responses for users depending on their extensive Google profile. Click on a major news site, and different people get different headlines and stories depending upon their demographics. Former Google CEO Eric Schmidt noted that individual targeting is 'so good it will be very hard for people to watch or consume something that has not in some sense been tailored to them.' The age of people sharing a similar digital experience is fading, with all that loss suggests."[15] This "me-oriented" system is thought to limit exposure to different worldviews while creating less empathy and understanding for people of different races, cultures, or perspectives.[16] This fragmentation would ultimately harm the creation of a potential online public sphere, which for any "new" communication medium would be a necessity, as it constitutes an important underlying tenet of democracy.[17]

Expanding on the public sphere and its relationship with democracy, scholars are also interested in the internet's relationship to power, class, and economic inequities. According to Dan Schiller, professor in the Department of Communication at the University of Illinois, "For decades we were told that ICTs (Information and Communication Technology) constitute a general source of economic uplift. From the theory of postindustrial society, first advanced during the 1960s, to 'new economy' boosters during the 1990s, to their successors today, public discourse has ordained the regenerative benefits of ICTs."[18] Additional theories during the 1960s and 1970s posited that growth in the information and technological sectors would also be able to transcend traditional inequities and class relations.[19]

The irony with these perspectives is that the mainstream economic premise that information should be treated as a commodity is ultimately flawed because information does little to fulfill its own definitional requirements of a commodity.[20] Signs and symbols should not always take the form of a price, and accessing the definition of information may not incur a cost. This internal inconsistency in turn questions the validity of mainstream economic approaches that conceptualize information as a commodity. Information can be subjective, shared, but it should also help provide meaning to material reality. There are important questions about how the information economy contributes to the maintenance of participatory democracy, especially for workers, consumers, and community participants.[21] Information is often ill-suited for commodity treatment, especially since access to it is necessary for democratic governance.

Other scholars, like Dan Schiller, Nick Dyer-Witheford, George Ritzer, and Nathan Jurgenson, regard the capitalist mode of production as subsuming information and communication technologies, which reinforce existing class and social relations.[22] This increasing inequality in turn harms democracy.[23] As long as profit-oriented corporations control important aspects of the internet, their main concern will be generating surplus, not democratizing the world.[24] They accumulate too much wealth and power by gathering information about all forms of human experience, which harms society and influences increasing nationalism, social divisions, and other threats to humanity.[25] This surveillance is also used for the social control of the poor and communities of color in the United States by reinforcing oppressive social relationships while enacting news modes of racial profiling.[26]

Moreover, other scholars view the internet, especially through social media, as exploiting online users by extending Marxian interpretations of labor to the digital world.[27] Here, the appropriation and commodification of digital labor not only produces economic growth but it also continues overall class divisions and asymmetrical power dynamics rooted in capitalism. These scholars emphasize that commercialization, exploitation, and economic division are not recent developments but predate the digital era. Theoretically positioning digital developments as part of the historical trajectory associated with a capitalist mode of production informs and guides the analyses presented in this chapter.

Online Advertising and Consumption Explosion

As chapter 3 noted, advertising became subsumed under capitalism to encourage consumption by creating demand for new items and markets. It helped create a consumer society by matching consumption to production starting with the transition into industrialization. Industries use advertising to develop a number of scientific techniques to manipulate desires, tastes, purchases, and consumption decisions, in order to maximize consumer capital extraction. The postindustrial, digital era significantly expands advertising and the use of data for sales purposes. It subsumed the internet once it became privatized in 1996, which changed its direction and purpose away from a public and noncommercial infrastructure to a new arrangement that favored surveillance in order to best modify human behavior to increase consumption.

The 1996 US Telecommunications Act, which helps govern telecommunications, internet access, and information services, also serves as a model for liberalizing telecommunication markets worldwide.[28] The liberalization of the internet assists in the development of online advertising, retail sales, and consumption. It also helps cultivate an explosively profitable industry in data mining, the purpose of which is to better persuade and influence individual online purchase decisions.[29] As the digital era continues to develop to a point where the extraction and sale of personal data for advertising purposes becomes a defining characteristic, individualized content, advertisements, and prices are made possible by increased commercial surveillance. This individualized online content, based on the surveillance of online users through mobile phones, computers, and tablets, can more effectively target and engage consumers than print and broadcast media.

While most capitalist economies are stagnant, it is clear that advertisers are accumulating significant profits during the digital era. Online advertising is expanding at rates that would make any Wall Street investor jealous, and much of the expansion is propelled by social media. As technology, software, and algorithms continue to advance, the individualization of advertisements, content, and prices also evolve to be more sophisticated and effective.[30] Price customization grew from these developments and should be viewed as an extension of contemporary advertising capabilities. Advertising in the digital age is a form of elevated corporate surveillance.

At this stage, with advertising growth rates through the roof, digital advertising is expected to become the dominant ad form in the world. As the fastest growing global media segment, total spending on digital advertising grew from $59.619 million in 2008 to $103.806 million by 2012 and was expected to reach $248.661 million by 2018. It grew by a 15.6 percent compound annual growth rate from 2008 to 2013 and was expected to increase at double-digit rates, growing 15.1 percent compounded annually to 2018 and account for 65 percent of the total increase in global advertising.[31] In 2008, its share compared to television, newspapers, and other media constituted only 14.6 percent, but by 2012 it had increased to 23.6 percent and was expected to constitute 38 percent by 2018. Global mobile advertising rose 83.9 percent in 2013, and was expected to increase by a projected 34.5 percent compounded annually over the next several years. According to the World Federation of Advertisers, US digital advertising reached $51 billion (out of $700 billion globally) in 2014, which was more than was going to all print media.[32] By 2018, it was expected to increase to $82 billion annually.[33] Moreover, well over $500 billion in revenue was projected to go to global advertising media owners in 2015, with over $150 billion going to the United States.[34] In total, the United States alone spends more than a trillion dollars a year on the sales effort.[35]

The growth of targeted, data-driven advertising in the digital/postindustrial era parallels a significant increase in consumption expenditures and e-commerce. According to the World Bank, US final consumption expenditure increased threefold, from $4.5 trillion in 1990 to almost $15 trillion in 2015, while global final consumption expenditure more than tripled, from $15 trillion in 1990 to $58 trillion by 2014.[36] These figures consist of expenditures on the goods and services consumed by individuals or collective communities. As the world has more access to broadband, consumer spending, particularly in Latin America, Central and Western Europe, and the Middle East/Africa is expected to increase dramatically.[37]

Revenue generated from e-commerce in the United States is growing at a rate of 16 percent per year, which is more than three times the 5 percent growth rate in overall US retail sales.[38] Estimated US retail e-commerce sales as a percentage of total retail sales increased from about 3.5 percent in 2008 to roughly 9 percent by 2017.[39] This places US retail e-commerce sales at about $111.5 billion by the second quarter of 2017.[40] Worldwide, global e-retail sales in 2016 grew by about 25 percent compared to 2015. The United States accounted for 22.2 percent of total global e-commerce retail, but it is projected to decrease to about 17 percent by 2020 as more countries increase their e-commerce. For example, the United Kingdom, China, South Korea, Germany, and Indonesia have some of the highest online shopping levels in the world.[41] These figures support Dan Schiller, who noted that digital information and communications have clearly created growth, akin to the nascent consumer industries of the 1930s.[42] Through commercial surveillance, they also helped propel increases in consumption.

Commercial Surveillance

Advertising and revenue increases form a close relationship with commercial surveillance, as the surveillance function enables the individualized advertisement to exist. New technology mixed with the neoliberal state and the complete commercialization of the internet created a massive online and offline surveillance function for private commercial sales efforts.[43] The surveillance of online consumers has become one of the fastest-growing businesses, even while the world's leading tech companies fail to offer basic disclosures about privacy to their users.[44] As of 2012, the average visit to a web page triggered fifty-six instances of data collection, and that number is most likely significantly higher today.[45] Marketers and advertisers constitute an important sector that challenges privacy and security. Significant sources of revenue come from marketing information about subscribers and customers to advertisers, who then decide how and what prices to advertise to these people. Scholars John Foster and Robert McChesney, as well as Vincent Mosco, originally labeled this surveillance capitalism, and it has its contradictions and hypocracies.[46] According to professor Christian Fuchs of the Westminster School of Media and Communication, "Capitalism protects privacy for the rich and companies, but, at the same time, legitimates violations of consumers' and citizens' privacy."[47] Moreover, it can violate basic civil liberties.[48]

There are several companies that surveil citizens, individualize online content, and customize online prices. For example, Acxiom's InfoBase®, the nation's largest repository of customer intelligence, also has the world's largest commercial consumer database, containing information on 500 million active consumers worldwide. It includes a majority of US adults and processes more than 50 trillion data transactions a year with about 1,500 data points per person.[49] Acxiom helps clients decide the value of people in terms of want or waste, and then determines the sales approach, which can cluster consumers based on an income spectrum to help determine product pricing. Their website notes "Acxiom's data and technology have transformed marketing—giving our clients the power to successfully manage audiences, personalize customer experiences and create profitable customer relationships."[50] Cambridge Analytica, the UK consulting firm owned by billionaire Donald Trump backer Robert Mercer, also has an immense amount of data on US citizens. The firm's CEO, Alexander Nix, claimed to have around four to five thousand data points on every adult in the United States.[51] Moreover, Facebook had an additional project that examined data from 57 billion friendships in 2013.

Another company called Next Jump gathers credit-card transaction data from American Express and MasterCard to infer what a person would be likely to purchase, and at what price. Next Jump's network includes 28,000 retailers that specify customer age, location, and income, while its software creates specific offerings to segmented consumers.[52] Another company called RapLeaf, which was purchased by TowerData in 2013, builds databases of individual voter-registration files, shopping histories, social-networking activities, and real estate records.[53] It has data on over 400 million consumers with over 900 million records and information on 52 billion friend connections.

There are several other companies that enable the surveillance of online users, examine creditworthiness, access public records, census data, tax assessments, telephone directories, and county records. They also have the ability to find individual zip codes, age, race, income, gender, degree of education, television and internet viewing habits, geographical information, social information, and other statistics to gather immense data on citizens in order to create individualized marketing messages and customized prices. Some of these companies include BlueKai, TowerData, Invidi, eXelate, Next Jump, Acxiom, Media6Degrees, Mindset Marketing, Experian, Equifax, TransUnion, Reed Elsevier, [24]7, RichRelevance, and Omniture. Additional platforms, such as AppNexus, Microsoft Ad Exchange, AOL's Marketplace, OpenX, Rubicon Project Exchange, and Smaato, use sophisticated pricing algorithms where prices depend not only on volumes but on advertisers' and publishers' profiles.[54]

Technologies and Tactics

This section describes additional technologies and tactics that many of the previously mentioned corporations use for surveillance. Some of the most important mechanisms consist of various forms of price customization, face recognition, and tracking software.

Price Customization Software (PCS)

Even though US retail sales topped $4.53 trillion, led economic growth, and represented 27 percent of nominal US GDP in 2013,[55] advertisers, marketers, and new technology companies realized there was still more money to be made by expanding the abilities of advertisers and the online sales effort. Perhaps the most efficient way of doing this is by collecting data through surveillance and online price customization, which are major growth areas for corporate profits.[56] Data mining can increase operating margins by 60 percent,[57] while the overall financial benefits of exploiting data, especially personal data within existing databases, is estimated to be around $600 billion.[58] The use of all kinds of consumer data, from income, to the brands of phones and computers people use, are then used to individuate prices for numerous online commodities and services.

With the advent of new information technologies, charging different online prices for different people, otherwise known as price customization, is regarded as a growth area,[59] falls within privacy regulations, and is becoming the evolving norm in the business.[60] Also referred to as personalized pricing, it constitutes the charging of different prices to end consumers based on a discriminatory variable, or, more simply, the ability to charge different prices to different consumers.[61] As a concept, price customization is analogous to price discrimination,[62] and can occur in at least three ways:

> Price discrimination is a logical corollary to behavioral targeting. Economists commonly identify three types of bias. First-degree price discrimination occurs when a different charge is tailored to a specific buyer based on what the seller knows about the customer. With the second-degree type, sellers openly offer a variety of fee options—for example, grocery discounts for buying large quantities or lowered bank fees for keeping large account balances—to induce consumers to choose the one that matches their interests or abilities

> to pay. In third-degree price discrimination, the seller decides what segments of the market have different levels of price sensitivity and charges the groups accordingly. Examples of third degree price discrimination are senior-citizen and student discounts.[63]

Price discrimination is similar to dynamic pricing, or behavior-based price discrimination, which can be defined as a simple form of price discrimination that offers different prices to different customers based on their previous purchasing history.[64] Dynamic prices can also be based around a number of factors such as geography, time, and temperature. *Perfect price discrimination* is a pricing strategy where each consumer is charged the maximum amount that they are willing to pay per item.[65] This is also called the "pain point," and in an "ideal" sales world, companies would price every product to each individual's pain point. In theory, price discrimination allows the firm to extract full surplus.[66]

Price customization works by performing "algorithmic profiling," which is a broader aggregation of data that allows companies to discriminate against consumers by putting them into specific groups slotted for marketing categories.[67] These profiling techniques can use credit checks and other mechanisms to customize rates and prices. According to one expert, "building up profiles over time allows firms to assess consumers' price sensitivity and then to start marketing campaigns either to individuals, or more likely, to segments that have varying degrees of willingness-to-pay."[68] Some of the variables that algorithms analyze to create customized prices include the kind of operating system/browser, the account on the site, and the history of clicked and purchased products.[69] Clicks, mouse overs, swipes, voice commands, cookies, and beacons are also used to track people in order to further gather data. "Fingerprinting" devices enable the personalization of mobile devices to track across media platforms. This results from consumers' abilities to delete cookies and the challenges mobile devices face for cookies as well. Furthermore, online ad networks, data exchanges, and data providers help boost advertising revenues by providing even more information about individuals.[70]

Companies create assessments of whether "an individual should be considered a desirable or an undesirable customer, whether a person would receive an ad for a particular product, what the ad should look like, how much the product should cost, and even what the copy should say," according to professor Joseph Turow.[71] This process exploits as much consumer data as possible while working to ensure that customers are separated and unaware that the prices and deals differ. Price customization is often premised on the idea that a consumer's willingness to pay exceeds the market price. However, these markets are not necessarily competitive and consumer choice decreases under these conditions. Many items that are price customized also entail limited arbitrage, or the ability to resell a purchased item for a higher price. This is particularly relevant with items that have restrictions, like airplane tickets.

The macro effect of price customization maximizes corporate profits, often between 8 percent and 25 percent more, and is considered superior to other forms of consumer capital extraction.[72] For example, big data and price customization have helped insurers sell at prices that have produced the maximum lifetime profit per customer to the tune of hundreds of millions of extra dollars.[73] One study from Stanford University found that advertisements that can individually target on the basis of ad viewing alone make up about a 16 percent increase in total potential profits, and that if an advertiser targets specific consumers, profits are higher under all ad and price targeting scenarios considered.[74] Another study by an economics professor at Brandeis University found Netflix could nominally increase its profits by using demographic data to adjust individualized prices based on income, as well as race, age, geographic location, and family size. If it used another 5,000 web browsing variables, like time on the internet (in particular during the day on Tuesdays and Thursdays) and if they visit Wikipedia, Netflix could increase its profits by 12.2 percent.[75] When these variables are implemented, the consumer estimated to have the "the highest value for Netflix would face prices about 40 percent higher when prices are tailored compared to when they are not," according to the study.[76] Unfortunately, companies do not disclose working details about PCS easily, and locating recent data on its overall economic impact is even more difficult.[77] Nonetheless, one estimate by Jupiter Communications placed the value of price customized e-commerce transactions at $7 billion in 2004.[78]

It is important to note that these trends are happening without public input, participation, or consent. A recent survey noted that 68 percent of respondents did not like individually targeted ads, while another poll indicated respondents did not want to be tracked online.[79] According to Turow, 62 percent of respondents thought it was illegal for online stores to charge different prices to various people at the same time of the day.[80] In fact, evidence suggests that when people realize that price discrimination is legal, they become angry at the businesses and the government for making it legal. They simply don't approve.[81]

This is interesting because consumers may accept spatial differences, where prices for the same item in an identical chain store may differ depending on the region or demand, among other influences. Consumers may also accept temporal differences, where prices may vary depending on the time of day or year, or according to demand, or a sale. Consumers may also accept bidding sites like eBay where prices are constantly in flux. However, with spatial differences there may be increased costs affiliated with the transportation of the items. With time-based price differences, customers may realize prices can change depending on the season, available floor space, or demand. With eBay, participants realize they are partially setting the prices among themselves by bidding.

But why should consumers reject different online prices for the same good or service? The main difference between these perspectives is that spatial and temporal differences affect all potential consumers equally during a certain time period. Ebay and similar sites enable consumers to compete among themselves while having some say in the price. These are all traits that are external to a person's identity. Price customization is completely different in that it treats individuals unequally and therefore discriminates based on personal information. This may partially explain why surveys demonstrate consumer misgivings and contempt.

Industries Using PCS

An increasing number of industries are using price customization software. It occurs in the travel, textbook, and pharmaceutical industries by charging more for tickets, books, and drugs, particularly in the overdeveloped wealthier countries. One study notes it frequently happens online within at least four general retailers and five travel sites, sometimes to the tune of several hundred dollar differences depending on the consumer.[82] Another study that examined 200 stores, such as Amazon and Staples, found prices can vary as much as 166 percent.[83] Andrew Fano, an Accenture consultant, notes that at least six of America's 10 biggest web retailers customize, but it is very difficult for shoppers to spot when it happens.[84] Cheaptickets, Orbitz, Expedia, Hotels.com, Home Depot, Staples, and Travelocity have practiced, or continue to practice, price customization.

For example, Staples utilizes online price customization by estimating consumer locations and distances from actual stores.[85] If there are competing stores close to the online consumer's location, the business would be less likely to increase online pricing. Lexis-Nexis, which provides data services and information, sells to nearly every user at a different price, while Victoria's Secret and other firms create different catalog prices based on average income differences across zip codes.[86]

Airlines and travel sites are also expanding their use of price customization. According to the International Air Transport Association, by mid-2015, 24 airlines were participating in price customization, or had announced their intention to participate. The International Air Transport Association prefers to call price customization new distribution capability schemas.[87] Travel sites are very likely to use price customization, and often automatic teller machine users are given higher prices than non-ATM users.[88]

Face Recognition

Besides price customization software, stores are also using facial recognition software to identify customers with previous or potential higher spending.[89] Face recognition is often associated with behavioral marketing. The technology works by analyzing video of people's faces as they enter a business, taking measurements to create a numerical code known as a "face template," and then by checking it against a database. This software is useful for identifying celebrities, wealthy customers, or other individuals who can afford high-end products. If it is a match, the program sends a message to one of the salespeople providing specific information, like the purchase history or sizes for the recognized person. These customers can receive special treatment and individualized prices.

Similarly, the NameTag app allows users to scan photographs of strangers and then see information about them, like their occupations or social-network profiles. This information can then be used to determine the extent to which prices could be set. With the FaceFirst Retail Facial Recognition software, one of the goals of the software is to "keep repeat customers coming back and spending more."[90] This software, now used in eight countries around the world, was originally designed to aid law enforcement, but is now used to aid the sales effort as it informs sales people exactly how much money customers in the database spend.[91] Facedeals is another example. It enables Facebook users to be offered special deals in retail stores that use the facial recognition software. It can target shoppers with in-store individualized advertisements, as well as collect customer behavior and integrate it with loyalty programs. The use of this software can increase bottom-line business profits.[92]

Tracking, Advertisements, and Prices

Many stores are using other technologies and digital tactics to track, assess, and create prices for consumers. The retail store Nordstrom at one point tracked its customers through Wi-Fi signals on their smartphones,[93] while food manufacturer Mondelez International is working on "smart shelves" that can show targeted messages to specific people and demographics when they look at goods on display. If the system identifies a middle-aged man, it will offer a QR code discount voucher, while for a teenage girl who picks up a soft drink, it will display the message that the drink only contains one calorie.[94] Facebook also partnered with a company called Datalogix, which tracks what people purchase in offline physical stores. Digital ads in physical spaces can also change with tracking information that automatically determines consumer likes and spending abilities.

These digital advertisements are created by the Microsoft Corporation. Their sensors can include

> capability for image processing, audio processing, light sensing, velocity sensing, direction sensing, proximity sensing, face recognition, pose recognition, transaction recognition, and biometric sensing, for example. A customer component analyzes the information and generates a profile about the customer. Advertisements are selected for presentation that target the customers as they walk in proximity of a presentation system of the store.[95]

These sensors are also used by law enforcement, in sports stadiums, retail outlets, and churches, with the latter being used to identify regular attendees to ask for financial donations.[96]

It is important to know that individualized content and advertisements are not limited to sensors, stores, Google, or social media. TV ad markets also allow advertisers to target advertisements to specific consumers based on the corporate surveillance of their previous product purchase patterns and their ad viewing behavior.[97] Digital streaming and set-top box technology now make televisions individually addressable through IP addresses or unique set-top box indentifiers.[98] Companies like Nielsen Catalina Solutions do this by merging credit card data from shopper loyalty cards with TV viewership data in order to provide advertisers and networks more detailed behavior-based profiles for viewers to purchase consumer packaged goods.[99] In addition, many marketers have examined emotional states closely and now target people for monetary extraction when they are considered the most emotionally vulnerable, which is Monday mornings for women.[100]

Online Labor and Consumer Exploitation

This section discusses areas where exploitation can occur in the digital era. Exploitation is inherent to capitalism and occurs when the group with the ownership of production exploits their workers by selling or trading the manufactured commodity at a level higher than the workers' paid wage.[101] There is little doubt that those who develop surveillance technology are exploited, as are the laborers who use the surveillance technology to gather and sell online user data. However, scholars disagree whether online labor sold in the form of data is to be considered coerced, alienated, or exploited.[102] This chapter considers "online free labor" to be exploited, and is therefore influenced by and aligned with scholars Christian Fuchs's and Mark Andrejevic's efforts to highlight how online user exploitation occurs.[103] This section identifies additional areas where online exploitation emerges. While Fuchs and Andrejevic largely focus on how online user labor is exploited, much of this section focuses on how exploited online user labor is then used to determine individualized online prices, which can further exploit consumers at the point of purchase.

Digital exploitation takes place through the collection and commodification of online user data by corporations like Google and Facebook. The data is subsequently sold to advertisers or other businesses for profit. This means the more users do online, the more revenue search engines and social media make as a result of this hidden system.[104] In essence, the basic strategy of the social media economy is to "crowdsource" value production to unpaid users.[105] The user labor involved in establishing consumer data is not adequately compensated and is thus exploited. This is also relevant to note as most profit-oriented social media, such as Facebook and YouTube, are almost entirely advertising financed.[106] Users seriously underestimate the economic value of the data they share with Google in exchange for accessing their services. In essence, Google exploits user ignorance to turn a profit.[107]

Exploitation can also occur at the purchase stage for consumers who buy an online item or service with an individually customized price. Similar to Andrejevic, who suggested value-generation processes should be expanded to consider the "role played by user data,"[108] value via pricing is now determined based on the exploitation of online user data generated through surveillance. This is essentially a digital version of Marx's "transformation problem,"[109] which is the conversion of values into prices of production.[110] While Marx was interested in labor and not pricing, the method and amount of capital extracted from online labor is unique within digital settings, as unpaid online labor can now determine customized pricing. It is a relatively new development, as Marx viewed the exploitation of labor as occurring "before the selling of the commodities."[111] The digital transformation problem appears to significantly enhance online surplus generation compared to offline sales efforts.

Digital price determination processes can be regarded as exploitative for at least three reasons. First, "exploitation is not simply about profit, but also about alienation," according to Andrejevic.[112] In this sense, consumers relinquish and lose control of their own productive capacity, not realizing the full value they have added to online commodity or service pricing. Exploitation of labor in this sense occurs when the surplus value created through labor is allocated without the consent of the workers. They do not get to choose who can purchase their data or what is done with it. Alienation ensues when their own participation has been turned against them as they become more and more dependent on technology that excludes them from key property that enables consumer agency. Consumers may end up having limited agency in how prices are set based on their labor. Second, the digital transformation problem also exploits because the online setting of individualized pricing is dependent on the appropriation of information extracted by online labor, which also represents a form of exploitation.[113] This information is extracted under an online structure of private ownership that turns information into a commodity. It is held captive to and conditioned by the perennially exploitative processes of capitalist exchange.[114] Third, another development based on user exploitation that can determine pricing is value cocreation, where the online consumer unknowingly helps create commodity value by contributing to its production.[115] Here, the seller enlists unpaid customers to channel their knowledge and taste preferences into product and marketing attributes, and then charges them for their own work. Commodity value becomes partially based on the consumer's exploited online labor.[116] Ultimately, the result of the digitally based price determination process founded on the exploitation of online user data through surveillance is seen as more effective at maximizing surplus value.[117]

Exacerbating Class and Economic Inequalities

In terms of exploitation and economic divides cultivated by the digital, advertising-driven online shopping infrastructure, initial examples indicate lower socioeconomic status (SES) consumers appear to be the most exploited, while those with the ability to consume the most, usually those with more capital, can receive discounts and perhaps pay less over time as a result of various loyalty or other programs that enhance consumption. In these cases, price customization can also benefit those who have more money. This is not surprising as 10 percent of US households account for nearly half of all consumer spending.[118] They are a valuable group to marketers.

At this stage, lower SES households appear to be paying more for the same online goods at some large retail chains than for higher SES households. According to a study from the *Wall Street Journal*, price customization software tended to create lower prices for higher SES areas, while providing higher prices for lower SES areas. In the study, the weighted average income of ZIP codes that mostly received discount prices was roughly $59,900, while the ZIP codes that saw generally high prices was $48,700.[119] Moreover, Nathan Newman, director of Data Justice, suggests that "online profiling based on user data allows seedier companies, from subprime mortgage lenders to payday lenders, to target the most naïve and vulnerable potential customers and facilitates new forms of price discrimination even by more legitimate firms that allows those companies to extract the highest potential price for goods and services from each customer. The result is harm to those victimized consumers and a more pervasive increase in economic inequality."[120] While PCS has also enabled the insurance industry to extract immense profits per customer,[121] consumer goods can often, but may not always, follow the same pattern as these industries.

While the logic behind PCS would also enable it to target the wealthy, by pricing online items much higher once the user data indicates the shopper has more money, this many not be the case. According to an interview with Nathan Newman:

> There is no reason that data analysis should lead to the rich paying more . . . the wealthy usually have more buying options so they are often charged less. . . . Look at credit, for example; the wealthy are usually offered lower interest rates and better perks on their credit

> cards, since banks may make less on each transaction but make more profit given a higher volume of total transactions. The cheapest rooms in Vegas are available to the wealthiest gamblers who are usually given free rooms to encourage them to come and hopefully lose money at the tables.[122]

Expanding on Newman's comments, professor Justin Lewis notes, "once a redistributive philosophy became unfashionable, the machinery of consumer capitalism—which is not built for altruistic purposes—was set to push its productive energy towards the 'haves' rather than the 'have nots.' Its mechanisms were directed towards those with the most money, *not* towards those most in need."[123] Moreover, because there are few studies that have examined the role of gender, race, and other variables with online price customization, it is still unclear if price discrimination targeted against protected classes is legal.[124]

Historically contextualizing the relationship between price customization and consumer agency suggests this is nothing new; only the heightened extractive capabilities from the digital era are new. Price customization existed prior to the Industrial Revolution and mass production,[125] although price was mostly determined by costs, not necessarily the seller's perception of the buyer's capital, data, or location. However, for centuries sellers in markets around the world have set, and continue to set, different prices for products based on perceived buyer potential. Historically price discrimination could happen through bargaining between the buyer and seller. Even today in the nondigital world, when one purchases a car from a dealership, the price may vary depending on the seller's perception of the buyer. At the same time, the consumer also has some agency in terms of defining the terms of the sale. With the new digital advertising and sales frontier, the historic consumer ability to haggle is now limited, and consumer agency is decreased. There is no "bargain" tab or option where consumers can try to reduce the price. If they can't locate another area where the item is sold, and there is no guarantee that outlet doesn't use PCS, then consumers are mostly left with the option not to purchase. Nonetheless, a similar asymmetrical relationship also exists in retail stores and supermarkets where consumers usually do not know the markup or the profit and most likely do not have the ability to bargain.

This online elimination of the consumer's ability to bargain for goods priced by PCS echoes other corporate developments that tried to decrease consumer agency. For example, when the music industry invented mp3 music files, it also eliminated the consumer's ability to resell the product.[126] This was important to the music industry as it lost profits from secondhand sales of music. If there are no used mp3s for sale at cheaper rates than new mp3s, consumers would be forced to purchase "new" and more expensive mp3 files. While sharing copyrighted music has always been illegal, digital formats made them easier to share among consumers. If PCS becomes the norm, consumers will have comparatively fewer avenues to avoid it.

Besides a deterioration of consumer agency, digital strategies like price discrimination are harmful to consumer welfare and disadvantageous to buyers.[127] In essence, with a few exceptions, the system of online price customization means higher overall prices for consumers compared to any model where prices are competitive, openly displaced, and advertised.[128] Another study found that average prices in mass markets where price customization did not exist were lower than in markets with targeted online advertising.[129] The overall effect of price discrimination decreases total consumer surplus.[130]

This means consumer surplus is redistributed to those who own the means of production partially as a consequence of surveillance and pricing capacities. Andrejevic sees surveillance as part of this exploitative process because it further extends the social relations that enable the few to profit from the actions of the many, which in turn further exacerbates class relations. He notes, "given its importance to the commercial development of the online economy, surveillance has a central role to play in any account of its distinctive forms of exploitation."[131] Moreover, "a form of class stratification emerges in the distinction between those who control and operate the means of communication and sociality, and those who trade their personal information for access to them."[132] The more stratified a society becomes, the more difficult it is to establish or maintain democracy.

These trends suggest that the asymmetry of online information means that the exploitative processes listed above function best when consumers are unaware of what is happening to them. In order to work well, price discrimination must be fairly secretive and difficult for online consumers to detect.[133] Asymmetry occurs when consumers do not know that prices within multiple industries and online shopping sites differ for the same item depending on who is looking at it. It also happens when consumers do not know that they may be charged more for items than other people based on what data corporations and advertising firms have about them. In addition, when individualized advertising techniques are using personal tastes, backgrounds, and preferences to potentially manipulate consumers—and none of this is regulated to inform online consumers that this is occurring—then there is an information asymmetry that is economically advantageous to capital.

Nonetheless, this asymmetry is dependent on correctly functioning algorithms, that consumers will not decline the transaction, and that even if they knew their prices were higher, or possibly lower, that they agree to them. Some consumers may well be aware of how data collection and price customization work, but others may not. For those that do, they may resist the purchase or choose to buy the item for a number of reasons. Others may be completely oblivious to data collection and price customization. Moreover, for active consumers with more time, websites that surveil online retailers and compare prices may also assist buyers by finding the best price, even if they are or are not customized. There will always be contradictions and antagonisms, as markets may not always skew purchases in favor of the seller.

Oligopolistic and Monopoly Industries

Exploitation also occurs outside of online labor and sales environments. It occurs in oligopolistic and monopoly industries, where a lack of competition creates a disproportionate economic and market advantage. In the digital industries, for example, Google, which is owned by Alphabet Inc., controls 70 percent of the search engine market and 97 percent of the mobile search market. Amazon sells between 70–80 percent of online books, while Microsoft continues to be used on 90 percent of all computers.[134] In addition, iTunes and the iPad dominate their respective markets while eBay and Facebook enjoy monopolistic power as well. Of the big five tech companies, Alphabet, Inc. is described as an advertising company first and foremost and earns 88 percent of its revenue stream from advertising, while Facebook earns 97 percent.[135] Apple and Microsoft earn most of their revenue from selling items that capture data and advertise, while Amazon is an e-commerce company that earns most of its revenue from selling products.[136] Microsoft earns about 9 percent, or $7.65 billion, of its $85 billion revenue from advertising, while Amazon also earns about 9 percent, or $12.51 billion, of its $136 billion revenue in "web services."[137]

The online surveillance, data collection, and price discrimination practiced by these and other companies seem to function most efficiently under monopolistic and oligopolistic conditions. The monopoly status of social media firms like Facebook and search engines like Google means that consumers share their data at too low a price.[138] One industry-oriented estimate placed the value of personal data to advertisers as up to $5,000 per person per year,[139] which suggests the trade off for "free" online services appears to be in favor of the businesses. In many countries, monopoly firms like Google and Facebook do not have serious competitors, so they can collect the most information on their users and sell their data at significant rates. Other companies have difficulties competing for potential advertisers because they have not been able to acquire as much consumer data. According to Robert Reich, well-known liberal economist, new technology firms like Google and Apple are spending more money acquiring patents from other companies, and also on litigation and lobbying than on research and development. Google is now the number one lobbyist in Washington and their goal is to prevent competition.[140] According to McChesney, "the internet monopolists sit at the commanding heights of U.S. and world capitalism."[141]

Acquiring consumer data from these online corporations can give other oligopolistic industries, especially in the retail and service industries, additional advantage through acquiring more valuable marketing information, and if they are effective, more market power. According to NASDAQ: IAC Publishing's Investopedia, which considers itself the world's leading source of financial content on the web, oligopolistic market structures also exist in the following industries: cable television services, entertainment industries (music and film), airline industry, oil and gas, pharmaceuticals, computer and software industry, cellular phone services, smart phone and computer operating systems, aluminum and steel, mass media, and the auto industry.[142] The goal with these and other oligopolies is to keep their market domination while enjoying the benefits that these market structures offer through enhanced profits. Indeed, Americans are spending more than needed due to oligopolistic market structures. They are spending more than needed on banking, health insurance, food, and other services and items while innovation appears stifled.[143]

Advertising can also enable manufacturers to gain market power, which allows them to raise prices and increase profits.[144] Ben Bagdikian, former dean at the University of California, Berkeley Graduate School of Journalism, notes an increasing number of economists find that advertising is "a major instrument by which big firms keep prices artificially high . . . advertising is used by many industries to maintain their power in the economy."[145] He also found an estimate that the prices in heavily advertised industries are 15 percent higher when compared to truly competitive free enterprise. The increased prices within these monopolistic/oligopolistic industries have in the past cost society between $10 billion to $20 billion per year. In this sense, advertising significantly affects both the allocation of resources and the distribution of wealth.[146] These consolidated market structures, often aided by advertising, more efficiently extract surplus to those who own the means of production.

Cumulative Effect on Consumer Sovereignty

The capitalist digital mode of production clearly exploits online users, and the cumulative effect of the previously described layers of exploitation related to this mode of production helps cultivate an overall degradation of consumer sovereignty and agency. Online user exploitation expands and impacts those who participate in online commercial exchanges, which means online user exploitation influences online consumer exploitation. Private ownership, concentrated markets, and the use of data extracted from exploited online labor also redistribute consumer capital more effectively to online businesses. The digital era lessens abilities to know when, how, and why online data and price customization are being used, which harms consumer sovereignty. In addition, this era also questions online user abilities to influence, counter, or control how data is used and prices are formulated, which suggests consumer agency is decreased.

While most people are not forced or coerced to be online, being online is increasingly required of them in order to participate in society, especially in Western societies. Indeed, without the internet, people are cut off from jobs, government services, education, and health care.[147] Being cut off appears to exacerbate social, economic, and racial disparities, at least in the United States. However, as I argue in this chapter, being online also makes users vulnerable to exploitation at many points of contact.

A careful analysis of these initial uses suggests this era will likely widen economic divisions. The postindustrial, advertising-driven digital era no doubt extends capitalism's exploitative capacities by enabling capital to heighten surplus value extraction. This is similar to scholars Ritzer and Jurgenson, who find capitalism's trajectory continued online by extending into new spaces to exploit new sources of profit.[148] In this sense, for developments in the digital realm, capitalism further subordinates activities under the law of value.[149] Or, as Schiller points out, the role of information and communications needs to be squarely placed within the political economy's chief developmental process.[150]

In terms of consumer interactions, the commercialized internet can empower users as Negroponte and Benkler believed it would, but it does not appear to be shifting control to consumers.[151] In certain instances it does give consumers more freedom of choice,[152] yet at the same time its infrastructure creates environments where consumers have fewer choices due to noncompetitive market structures and pricing scenarios. Moreover, many note the surveillance function of the digital era further reinforces oppressive relationships, especially among lower SES and minority communities.[153]

While the use of data, scores, and algorithms in the digital era has the potential to minimize or maximize discrimination and economic inequality, it appears they are more likely to maximize it.[154] Scholars note that the trends within the digital era can increase inequality and thus harm democracy.[155] Turow and Schiller were correct to note it would be unlikely for the internet to transcend traditional inequities and class relations.[156] However, when Turow suggested this new individualized advertising/sales system would reinforce preexisting inequities, he was only partially correct.[157] It appears to be widening them.

This chapter suggests that private ownership of the online information structure shapes the use of new media technologies in ways that reproduce and *enhance* existing power relations. To be more precise, the postindustrial, advertising-driven digital era highlights how initial results of price discrimination and online exploitation reproduce and further *exacerbate* economic inequalities. The next chapter discusses the implications of these trends as they contribute to the global warming emergency in many ways. Without reform, the existing digital infrastructure ultimately undermines efforts to address and mitigate the climate emergency.

CHAPTER SEVEN

Challenges to Sustainability in the Digital Era

The previous chapter found that the postindustrial, advertiser-driven digital era exploits online users while contributing to the upward redistribution of societal wealth. Its surveillance function aids in the enhancement of consumption by acquiring data on consumer habits and preferences, which then allows advertisers to better target potential consumers with highly informed persuasive messages. In this chapter I focus on how these trends contribute to the climate emergency. I argue that the present trajectory of the advertising-led digital era undermines efforts to address the climate emergency in at least five ways.

First, one of the aims of the advertising-driven digital era is to modify online user behavior to increase consumption, which counters science that tells us of the need to *decrease* overall consumption. To do this, the digital era's commercial online surveillance focus is used to create personalized messages that are often more effective in persuading online consumers to enhance their consumption habits compared to nondigital efforts. These developments parallel a dramatic increase in global consumption since the commercialization of the internet. Second, the production and disposal of electronic items that support online platforms, along with digital obsolescence, significantly pollutes the environment while unnecessarily draining the planet of natural resources. Both contribute to planetary warming and counter efforts to fight the climate crisis.

Third, the amount of energy to power the digital era is unsustainable. As Jennifer Rauch, associate professor at Long Island University in Brooklyn, notes, "From a sustainability perspective, growth is not a criterion for progress and proliferation of digital media cannot be limitless."[1] Powering data centers, cloud computing, and online advertisements, among other entities, has environmental implications that counter what is needed to address the climate emergency. Fourth, the digital era enhances surplus, enriching a small but significant percentage of global millionaires. This counters efforts to address the climate crisis because the wealthy consume more and thus contribute more to global warming by their overgeneration of carbon emissions. Fifth, the control over data collection and price customization indirectly harms democratic processes by exacerbating economic inequalities. The extensive surplus generated from the digital era further enhances industry power while negatively affecting economic inequities. These trends harm democratic processes that are essential in addressing the climate emergency. Addressing economic inequality while democratizing political systems will better help address the climate crisis. The sum of these five trends challenges efforts to combat the climate emergency and indicates that the postindustrial digital era is ultimately unsustainable for the planet.

Increasing Consumption

The idea that the new technologies of the digital era would be able to reduce stress on the environment by promoting dematerialization of production and consumption advocated by some scholars has yet to turn into a reality.[2] In fact, the opposite is occurring. One of the advertising-driven digital era's main goals is behavioral modification to enhance consumption. To address the climate crisis, overall consumption needs to immediately decline to reduce stress on the environment. Instead, as the previous chapter noted, digital advertising growth rates are the fastest growing global media segment, and their goal is to create more and more demand for products and services. The digital era's commercial online surveillance focus is used to create an overwhelming number of personalized messages that try to better persuade online consumers to enhance their consumption habits. These messages more effectively target and engage consumers compared to print and broadcast media. They also replace and overpower messages that educate consumers about the global climate emergency and how humanity can work to mitigate it. Their growth parallels a tripling of global final consumption expenditure during the past 30 years, with overall sales and e-retail sales having increased significantly.

The net effect of the postindustrial digital era, through the use of the internet and new technologies, is leading the world to greater consumption and resource depletion.[3] The global consumption of new goods and services "has increased dramatically over the last decades, in both absolute and per capita terms," according to the IPCC.[4] This is extremely problematic because the production processes of these goods are responsible for the majority of environmentally destructive emissions,[5] while the consumption of these goods is a key driver of environmental degradation, including global warming.[6]

As previously noted, individual consumption is responsible for 64 percent of global emissions, while the remaining 36 percent are attributed to consumption by governments through investments in infrastructure and international transport.[7] Indeed, US residents are increasing their consumption habits.[8] It is quite clear that the digital era contributes to increasing consumption and energy use.

Production and Disposal of Electronics, and Digital Obsolescence

The production and disposal of electronic devices also harms nature and humans, while the twofold business strategy of progressive or planned obsolescence also damages the environment by unnecessarily draining natural resources. These digital era trends negatively affect efforts to address the climate emergency.

The production and disposal of digital devices by which advertisers target consumers in the digital era (meaning computers, cell phones, tablets, and other items) do significant environmental damage.[9] In terms of production, the amount of resources it takes to build electronic items often surpasses the rate at which they are used. Electronic items can have a high resource intensity. For example, flat-screen televisions and computers require four times as much energy to make than is expended consuming them.[10] In addition, by 2002, a desktop computer and monitor required 50 pounds of chemicals, 3,300 pounds of water, and 530 pounds of fossil fuels to produce, which is roughly the equivalent of the resources used to manufacture a large automobile.[11]

Moreover, when digital communication technologies are thrown away their afterlives continue in the form of e-waste, which contain numerous metals and toxic chemicals that harm humans, other living things, and the overall environment.[12] E-waste is one of the fastest growing elements of municipal cleanups around the Global North. About 44.7 million metric tons was generated in 2016, with most of it emanating from North America, Europe, and other high technological turnover areas.[13] Much of it comes from televisions, computers, and cell phones.[14] This waste is then often illegally shipped to Africa, Latin America, and the Far East for disposal, where it causes pollution and additional health hazards.[15]

The manufacturers of electronic products in the digital era also practice the old "analog-era" strategy of progressive/planned obsolescence, which unnecessarily drains natural resources. It predates the digital era by several decades and complements economic policies that prioritize growth and capital gain over natural resource sustainability. However, it takes on a slightly different mode in the digital era. The first aspect of obsolescence is the advertising and marketing strategy. The goal here is to use marketing and advertising campaigns to try and convince people to discard their already functional items for a similar, yet slightly "newer and improved" product. The aim of this persuasive technique is to shorten product lifespans in peoples' minds and convince them that replacing items is good for the economy. In this sense, Apple spends hundreds of millions of dollars to advertise newer versions of goods that people already have.[16]

The second aspect of obsolescence is the business strategy. Here, new objects are designed to break down after only a few years. This process is purposefully designed by engineers to maximize potential revenue.[17] Not surprisingly, it is often more profitable to design an item with cheaper components to break down after a few years instead of building a product to last for several decades.[18] As one analyst from the 1950s noted, consumer goods are best consumed, burned up, worn out, replaced, or discarded at ever increasing rates.[19] This makes perfect business sense, but not environmental sense. If a product was purposefully made extremely well to last a long time and to be less likely to break down, it is likely consumers would not need to purchase another similar product in any rush. However, that means that the company producing the product will not make as much money as the industry where products are meant to break down after a few years.

Another example of obsolescence exists when the products work fine after a time, but are no longer useful. This is one of the key features of progressive obsolescence in the digital era. The electronics and computer industries are at the forefront of this strategy. For example, even though cell phones work fine regarding phone calls and domestic texts, consumers often have to replace them with a newer model in order to text international friends, use recent apps, and use it when traveling outside of the United States. Apple does something similar by limiting its "older" phones from being able to run newer software, apps, or other features. Old phones often work fine, but because of digital locks and keys engineered to allow or prevent access,[20] many functions are rendered useless. This means that even if someone's phone works, they will have to buy a new one in order to access up-to-date features. About 450,000 cell phones are discarded every day in the United States.[21]

What is the effect of obsolescence on the environment? With electronics, most of them are thrown out and into incinerators and dumps, and even when converted to waste, they are highly toxic. One calculation has the annual turnover rate at about 17 percent of electronic products,[22] which is the percentage per year that consumers are getting rid of their electronic items. Indeed, hundreds of millions of still working items that constitute billions of pounds of hazardous waste are thrown away every year.[23] The business strategy of forcing consumers to continually upgrade electronic gadgets by making many of their functions obsolete unnecessarily expends natural resources.

Unsustainable Energy Use

Besides the exhausting of natural resources, the postindustrial digital era also increases energy use. These are at least two areas that need to be decreased in order to address the climate crisis. The data centers that acquire consumer information to better target and advertise online, combined with the electronic devices that are surveilled, require a massive amount of energy to function. Service providers prefer as much online advertising as possible. The more online advertising, the more data is used, which means they earn more profit. This is why some general service contracts charge to access the internet based on the amount of data used. Internet service providers and advertising, in addition to consumption and profit, are important variables here.

The digital era's goal of increasing our unsustainable overconsumption of overproduced goods and services is unsustainable as it overuses and expends too many nonrenewable energy sources to promote our overconsumption. Its need to collect and render data about us through an ever-expanding list of digitally connected gadgets has as its main goal to increase human consumption of goods and services in order to increase commercial profits. The list of devices connected to the internet that collect our information is already quite large and will continue growing. Besides the internet itself, social media, eyeglasses, cars, phones, computers, watches, toilets, houses, internet kiosks in cities, apps, televisions, children's toys and dolls, the Echo, Alexa, and Google Home all collect our data.[24] It's uploaded onto servers in data centers to be stored and analyzed, which requires an unsustainable amount of energy to function.

Much of this information is then transformed into advertising messages. Heightened data use to help advertising also requires elevated levels of energy extraction to power data centers and cloud computing to create the advertisements in the first place. Advertising content in the digital world expends significant energy, consuming between 10 and 50 percent of user data plans, according to the ad-blocking software provider Shine. Ad blocking software was estimated to have cost publishers close to $22 billion during 2015.[25]

Vincent Mosco points out that the rates of electrical consumption for data centers and electronic devices are ultimately unsustainable.[26] Likewise, most energy feeding them is not renewable and the elevated levels of energy required contribute to GHG emissions. For example, the power demands of data centers grew by 56 percent between 2005 and 2010, and they continue to expand and require more energy.[27] By 2013, data centers used 2 percent of all of the electricity consumed in the world, while their carbon emissions were set to quadruple by 2020.[28] A single data center can use more power than a medium-sized town.[29] Moreover, the energy sources are often unsustainable. For example, Facebook built a data center in central Oregon that was serviced by a utility that primarily used coal-fired power stations, which are some of the largest sources of GHG emissions in the United States.[30] Most internet and cloud platform companies, along with colocation data centers, which are digital landlords that lease out large pieces of data center space to host the servers from companies like Uber, Facebook, Google, and others, are also powered by very little renewable energy. The percentage of renewable energy used in Virginia's Data Center Alley, which is home to the world's highest concentration of data centers, is only 4 percent. Data Center Alley claims that 70 percent of the world's internet traffic passes through its borders.[31]

Power hungry services like video streaming, which also run through data centers, appear to be increasing around the world, given that the global video streaming market size was valued at $38.56 billion in 2018 and is projected to massively increase to $149.43 billion by 2026.[32] Cloud computing is also expanding, and draining more energy as some 300 million people were reportedly using Apple's iCloud every day in April 2013.[33] Cloud computing has become a priority with Microsoft and their cloud business is growing quickly with profits increasing dramatically.[34] Apart from data centers, the phones, tablets, computers, and other electronic items, which add up to over 10 billion devices, use 15 percent of all global residential energy. They will require 30 percent of the world electrical grid by 2022, and 45 percent by 2030.[35] Currently, if cloud computing was a country, it would be the sixth largest consumer of electricity in the world.[36]

Expanding cloud computing, streaming, and other online services is not good because the data centers that enable these entities already have a larger carbon footprint than digital devices or the airline industry.[37] Overall, according to a study by Gartner Inc., the global information and communications industry accounted for roughly 2 percent of global CO_2 emissions in 2007, which was equivalent to the carbon-heavy aviation industry.[38] This was also a conservative estimate as the study did not include consumer electronics other than cell phones and personal computers. In many ways, this means that watching streaming video, gaming, and shopping online will be worse than flying, if it already isn't. It also means that all of those online activities, along with tons of others, help fuel the digital surveillance system to further promote and encourage the unsustainable overconsumption of overproduced goods.

In the best theoretical mitigation scenarios, global energy sector carbon dioxide emissions are supposed to decline, yet increases in energy use for electronic gadgets do not seem to be factored in, which questions these scenarios. For the expanding consumer base of these consumer products that "need" to be constantly thrown away, upgraded, and purchased anew, the digital era is clearly making it more difficult for the world to decrease carbon emissions, especially with the manufacturing and transporting of these goods. The transport sector's baseline carbon dioxide emissions are projected to double by 2050, while industry's emissions are projected to increase by 50–150 percent by the same year.[39] The US Energy Information Administration or EIA projects a 48 percent increase in world energy consumption by 2040.[40] This is extremely troubling given that "the most energy-intensive countries should reduce their current per capita energy consumption by 70% to allow the least energy-intensive ones to increase it by 30%," according to one group of experts.[41]

The largest corporations in the digital era consist of social media corporations that also use cloud computing. Not only do they overuse energy, but they also work in favor of the fossil fuel industry by profiting from their money and by promoting their propaganda. For example, corporations like Facebook and Instagram have no problem accepting large amounts of money from big oil to promote fossil fuel production through targeted advertisements on social media. They recently earned a few million dollars during the 2018 US midterm elections to promote one-sided industry information sent out to specific members of the public.[42] Accepting this fossil fuel money means social media corporations are willing participants in misinforming the public, which in turn slows down the potential for the collective action required to facilitate policy change. In this situation, social media corporations clearly sided with their personal profits.

Extending Economic Inequality and Increasing Larger Carbon Footprints

Besides expending unsustainable energy use, digital era corporations continue to pull in massive profits. The revenue generated during the digital era enriches a small segment of the population that consumes items requiring larger carbon footprints.[43] Increased numbers of wealthy people is a concern because economic inequities pose serious challenges for mitigating climate change as the rich consume the most and cause the highest carbon emissions.[44] Their numbers need to decrease, not increase. According to a study that compared GNP growth rates and consumption rates throughout OECD countries for several decades, income growth and consumption growth are interlinked.[45] This means that the more money one makes, the more she or he will consume, which translates to more harm done to the environment. As noted earlier, this suggests that focusing on income inequality, as one of many problematic variables that impact global warming, needs to be addressed in the struggle for a sustainable planet.

Comparing recent lists of the wealthiest industries and people to lists of thirty years ago highlights that many of today's wealthiest corporations, now in the digital industries, have only recently existed. Many of the newly rich come from technology corporations. They are cashing in, many from digital advertising, on the profits generated from the US government's decision to privatize and commercialize the internet. However, the transfer of wealth is not limited to corporations in the digital sectors and also extends to nondigital manufacturing and service sectors. The nondigital sectors use individualized data to better customize and advertise in hopes of increasing sales and profits. Surveillance, user data, PCS, and advertising *contribute* to the transfer of wealth to the group that owns the means of production, for the software that enables the surveillance ability, and to the makers of the commodities that will be sold through its function.

Of the world's wealthiest 200 people, based on their net worth, the number of billionaires who have accumulated their wealth from telecommunications, media, and technology is quite large, with 46 people in that group. Half of the world's wealthiest top 10 people were involved in this sector. Of the 46 wealthiest people, the majority came from the technology industry.[46] Of the largest US corporations from 2012, 13 of the top 30 were from internet and digital-related industries largely considered either oligopolistic or monopolistic.[47]

Much of this wealth came from public subsidies. States and cities in the United States, many with failing infrastructures, broken budgets, and struggling schools, have given some of the world's most profitable tech corporations—Apple, Amazon, Google, Facebook, and others—at least $9.3 billion in subsidies during a five-year period between 2013 and 2017.[48] Moreover, as the majority of the 500 largest US corporations keep $2.1 trillion in tax havens through tax loopholes created by Congress, about $620 billion in US taxes are lost to local, state, and federal governments, with ordinary citizens making up the difference.[49] Many of these corporations, like Microsoft and Apple, are some of the largest holders of offshore money not subject to US taxes, with Apple holding $181 billion and Microsoft with $108 billion.

Many of the people within the digital and technology industries can be considered part of the newly minted population of global millionaires. This population increased from 8.7 million in 2005 to 13 million in 2016, while the global population of ultra-high-net-worth individuals grew by 61 percent from 2005 to 2016 to constitute 187,500 people with $30 million or more in net assets, excluding their principal residence. These individuals have assets ranging from over one million to billions of dollars and hold net assets worth around US $66 trillion, which is more than the value of all global equities.[50] By 2019, an additional 31,000 people became ultra-high-net-worth individuals, and roughly half of this global population lives in the United States.[51] Globally, about 80 individuals had the same net wealth as 3.5 billion people, or half of the entire global population in 2015.[52]

Global millionaires are recently increasing their spending on luxury items, especially super-yachts, which are boats longer than 24 meters, or close to 80 feet. Sales of these boats rose by 40 percent in 2015.[53] Wealthy Americans also appear to be splurging with additional high-end consumption habits. Luxury automakers like Bentley and Porsche report rising sales, luxury hotels are seeing increased sales, while wealthy Americans are expected to continue increasing spending on luxury hotels and resorts (by 17%), home entertainment and electronics (by 17%), watches (by 10%), and automobiles (by 18%).[54] Spending on items and experiences that require elevated amounts of carbon emissions (like airfare) will continue to increase in the future as the number of ultra-high-net-worth individuals is expected to rise 41 percent to 263,500 by 2025, while the population of millionaires is expected to increase from 13 million in 2016 to 18 million by 2025.[55]

While it may seem at face value a good thing if economic policies increase the number of millionaires, even if that number is not very large at all compared to the overall percentage of nonmillionaires in society, an increase in millionaires is not a good thing for those who connect consumption to the environmental crisis. This is particularly important as population estimates put the planetary population at 9.7 billion inhabitants by 2050, up from 7.3 billion in 2016. An increase of 2.4 billion people will further strain the already overconsumed planet. Any system that increases the numbers of the wealthy naturally clashes with efforts to mitigate the climate crisis, not only through overconsumption of resources but also by impeding government revenue that could be used to adapt to and mitigate global warming.

Challenging Democratic Governance

Instead of trying to remediate social and economic inequities, the postindustrial, advertising-driven digital era is exploiting them. With information the capitalist currency of the online information economy, consumer data collection, advertising, and PCS further enable the redistribution of wealth. When companies, through data mining, big data, and individual behavior profiling, know more about consumers than those consumers know about their marketplace options, this information asymmetry contributes to overall social economic inequity.[56] Even according to a White House report, price discrimination "transfers value from consumers to shareholders, which generally leads to an increase in inequality."[57] The macro effect of intensifying economic inequalities gives those with more capital a disproportionate political influence, which harms democratic governance.[58] Exacerbating these socioeconomic inequities creates significant challenges for the democratic governance that is needed to address the climate emergency.

Systemic inequities existed in the offline world well before the internet, and the widening wealth gap should never be solely attributed to online activity. However, the evolution and expansion of online activity does not appear to narrow the continually expanding levels of economic inequality. Consumer exploitation and the corporate ability to extract increasingly higher sums of money online, particularly from lower SES peoples, is troubling, as the United States has one of the worst income and wealth distributions on the planet. This leads to increases in the small but significant number of very wealthy, while further enriching many who are already very wealthy.

The rise in income inequity in the United States is much greater than in other rich capitalist countries.[59] The enormity of these economic disparities is astounding and unprecedented, with the US middle class having a much smaller percent of national wealth compared to every other region of the world.[60] The top 10 percent of the US population owns 77.2 percent of wealth, the top 1 percent owns 41.8 percent, the top 0.1 percent owns 22 percent, while the top 0.01 percent own 11.2 percent of the nation's wealth.[61] The average mean net worth of the wealthiest 1 percent in 2010 was $16.4 million, while the average for the bottom 40 percent was in the negative at $-10,600.[62] A recent report for the Institute for Policy Studies found that America's wealthiest 20 people own more wealth than half the American population, which is about 152 million people.[63]

Roughly 48 percent of Americans are either poor or low income,[64] while the Urban Institute reported that half of all Americans (51%) experience poverty at some time before age 65.[65] A global wealth study by Credit Suisse found that of the world's poorest decile, or bottom 10 percent, one out of ten is from the United States, which means nearly 50 million are part of the world's poorest 10 percent.[66] In addition, almost a million people were estimated to be homeless on any given day in 2007, while about 3.5 million people were homeless at some point during that year.[67] In projecting to the future, economist Thomas Piketty predicts wealth will continue to become more concentrated at the top.[68]

Controlling asymmetric consumer information within the digital era enables those who own the means of production to more effectively extract surplus, which further divides class relations. Societies with enhanced class divisions are less democratic because those with more capital often have an unequal influence over power. Professor Robert McChesney notes that "inequity promotes a thoroughgoing corruption of the governing process as wealthy special interests come to dominate, forcing the system to maintain and even increase their privileges further."[69] These interests are quite pronounced and influential in the US political system. A total of 158 families, along with the companies they own or control, contributed nearly half of the seed money to support Democratic and Republican presidential candidates, with the vast majority of money going to the Republican Party.[70] This group receives significant political and media attention, even though the poor make up a much larger percentage of the US population.[71]

According to authors John Nichols and Robert McChesney, an important aspect of the democratic process has "as its purpose the redistribution of power from elites to the great mass of people."[72] Focusing on mechanisms from the digital system that exacerbate inequities becomes extremely important as the United States has slipped out of the list of fully functioning democracies during the past few years. According to the well-known Democracy Index from the conservative *Economist* magazine, the United States is no longer a fully democratic country. It has fallen off the top 20 list during the past decade and is now considered a flawed democracy.[73]

Efforts to address climate change have to go through political systems and many politicians are considered bought off by the wealthy given the nature of money in the political process. Much of this money comes from various industries that profit from the status quo. Therefore, any system that further enriches the class that largely rejects government regulation in favor of public initiatives, like mitigating the climate emergency, must be changed to one that doesn't extend this damaging trend.

Need for More Digital Era Solutions

This chapter suggests the digital era contributes to the climate emergency and that the amount of energy required to propel this system into the future is unsustainable. Therefore, more scholars and scientists will need to find new ways to reimagine a sustainable future for humanity's relationship with technology. To address the problem of the online infrastructure's contribution to exacerbating economic inequalities, one possible solution would be to lobby the Federal Trade Commission to stop price customization while breaking up monopolistic and oligopolistic markets. There is also a need to create a nonprofit and noncommercial internet where surveillance is not allowed. Creating laws that allow individuals to own their own data will also help.[74] These initiatives may help decrease surplus extraction toward the top, perhaps help address some US democratic deficiencies, and assist the path toward economic justice.

Jennifer Rauch's excellent book *Slow Media* offers a number of additional solutions to the eco-challenged digital world. She suggests media need to be sustainable, and defines the concept "as values and practices characteristic of a media culture able to indefinitely maintain biodiversity and a high quality of life while avoiding the long-term depletion of natural and human resources."[75] She proposes finding a middle ground where society should ask itself how much technology is really necessary in their lives. She also suggests "facilitating self-education and integrating green cultural citizenship into more high-school and college curricula."[76] In addition, manufacturers of digital gadgets should make things last longer and not obsolesce, make things easily repairable, while making information on their supply chains openly available. Groups like Greenpeace, GoodElectronics Network, and the International Campaign for Responsible Technology also advocate for a global industry standard that complies with sustainability standards and human rights. Last, rebooting Luddism to inspire creative new attitudes and behaviors toward the digital world, while working less, can also diminish our digital carbon footprints. One could add that creating alternatives to capitalism is urgently needed in order to better address increasing economic inequalities, democratic deficits, and other challenges facing humankind.

While collective efforts increase to address the climate emergency, the sources of energy and its uses will become more and more crucial. It seems there won't be enough energy for everything that humanity wants, or what commercial societies tell us we want, without significant additional harm being done to the planet. Will much of future renewable energy end up going to businesses, so their profits can continue at the expense of other forms of more harmful energy use going to other basic needs for humanity? Or will renewable energy be prioritized for basic human needs and human rights first, perhaps collectively decided? These are serious questions that society will need to discuss and debate quickly. However, as the next section notes, mediated discussions of such powerful and important topics are extremely challenged, particularly within US news media structures.

IV

Journalism and Climate Change

CHAPTER EIGHT

Challenges to US Climate Change Journalism

> People listen to you. They are influenced by you, and therefore, you have an enormous responsibility. And let's be honest, this is a responsibility that most of you have failed to take.[1]
>
> —Greta Thunberg

The last section focused on the digital era, its infrastructure and technologies, and their relationship with the climate emergency. This section is composed of three chapters that examine journalism and climate change. Chapter 8 looks at the major challenges facing US climate change journalism, chapter 9 continues this area of focus by examining fossil fuel industry misinformation, blogs, and their relationship with public opinion, while chapter 10 argues that media reform is necessary in order to lay the groundwork for significantly improving US climate change journalism.

This chapter discusses several, mostly US-based trends in commercial journalism, noting how they impact the quantity and quality of climate change reporting. These trends indicate journalists are under immense constraints, which negatively affect the kind of reporting and the output of news-media generated climate change information. The consequences of these limitations negatively affect democratic processes by limiting the public's access to global warming information and contributing to the delay of crucial political policy solutions needed to address the climate crisis.

I begin by noting how important the news media are regarding climate change information, given that they are the main source of public scientific information. Global warming is also a slow and often invisible process that demands accurate and informative reporting. Next, I outline a brief history of climate journalism and its relationship with climate denialism, noting how US commercial news media were slow to give global warming significant news coverage, even though climate change research has been around since the late eighteenth century. I then discuss how the total number of journalists has been declining since the great economic recession of the late 2000s, where the industry restructured itself by laying off a large number of reporters. The decrease in the number of journalists then went on to influence a significant decline in climate change coverage.

The chapter continues by analyzing how the journalistic norm of "balance" challenges climate change coverage by creating a false balance, which then leads to bias. Comparing coverage to most other industrialized Western countries, US mainstream news media coverage of climate change exhibits one of the worst track records because it traditionally has supported extreme denialist and skeptic minority views. The last two sections of this chapter focus on news media silence regarding climate change reporting. Editors and the influence of advertisers can be challenging for journalists interested in writing about climate change because most commercial news media organizations are profit-oriented and dependent on advertisers for funding. Climate change issues often clash with advertising interests, while stories about climate change can be depressing, which usually don't attract high ratings. Editors are aware of these pressures and, as a result, tend to focus on other topics that increase ratings, curry favor with advertising interests, and generate higher profits for the company. Last, these trends influence overall news media silence regarding climate change. This happens when commercial news media ignore climate change, especially by deciding not to connect extreme weather or other environmental phenomena with global warming.

Importance of News Media

The manner in which mass media, and in particular news media, affect public perceptions of scientific issues is quite important. The media are, by all means, the main source of information by which millions of readers and viewers receive much of their scientific information.[2] With most people having very little direct contact with global warming, as it is a slow and often invisible process, citizens are largely dependent on the mass media, not only for news about climate change but also for related topics and human activities that are considered responsible.[3] In this sense, public understanding of global warming is a mass communication problem.[4]

Media coverage of climate change directly affects the levels of public concern.[5] Coverage can influence what specific topics people are supposed to think about, which in turn can affect how science is translated into public policy. Addressing the climate crisis is a collective action problem and therefore requires the public to facilitate policy change.[6] It is important to remember that media coverage does not guarantee changes in behavior or immediate comprehension by media consumers. Nonetheless, it can influence how the public thinks and acts about a topic.

Decades of communication research demonstrate how the public thinks about an issue can often be determined by the media's framing of a specific issue.[7] Media framing occurs when some aspects of a perceived reality are selected over others. These aspects are made to appear more salient than others in such a way to promote a particular problem definition, casual interpretation, moral evaluation, or treatment recommendation for the aspect described.[8] Media framing of climate change is complex and can manifest itself though labels such as "consensus" or "controversial," "scientific," "unanimous," or "skeptical." Other descriptions such as "emergency," or "nonemergency" or "crisis" are also relevant. These frames in turn influence public perceptions of climate change.

Indeed, various sectors of society try to influence the manner in which news media communicate scientific information through their own framing of issues. As important as framing is to communication research on climate change, it is only one of many influences that factor into the creation of news coverage. As such, studying news media coverage of global warming should never be limited to a question of framing.[9] Structures, power relations, parameters, and values, which can be aligned with special interests, assist the media's construction of events.[10] Human decisions are made that include or exclude information as phenomena are interpreted. The press is an instrument of social power and should not be underestimated in its ability to impact society through the way it structures and shapes "reality."[11]

Indeed, the mass media constitute complex institutions and processes by which environmental meanings are produced and consumed,[12] and many political and economic interests see the news media as their main targets to affect public opinion. It is important for people to understand how news is created as "no serious scholar today regards the daily 'news' as an objectively discernable set of events that is collected and disseminated by news organs in a wholly unbiased manner," according to scholars Allan Mazur and Jinling Lee.[13]

Brief History of Climate Journalism and Climate Denialism

US commercial news media were slow to give global warming significant news coverage, even though research on climate change goes back as far as the eighteenth century.[14] The *New York Times* is cited as being one of the first major publications to cover climate change during the 1930s.[15] It went on to cover human contributions to climate change later in the 1950s by exploring links between global warming, technological processes, agricultural shifts, and sea level rise. Coverage died down for a few decades afterward, but resurfaced in the 1980s when the *New York Times* combined with a few other national media organizations to place global climate issues back on the American agenda.[16]

The increase in coverage became a concern for large industries, which resulted in new strategies to alter and influence coverage. One strategy entailed providing politicians with information questioning climate change. By the 1990s, government officials began to repeat denialist or skeptic rhetoric generated by industry-connected think tanks and public relations firms. Part of the increase in denialist rhetoric included within the anti-environmental countermovement also stems from the shift when Republicans controlled the US Congress. They enjoyed significant numbers, which enabled them to exploit news practices of "balance" and supply official sources to news media.[17] In turn, these officials at various points in time became the most cited sources, outgunning actual scientists in the larger prestige/agenda-setting press.[18] This was happening even as one study found denialists accounted for only about a dozen scientists.[19]

Eventually denialists preferred to be silent on global warming, or at least question or confuse the public with skeptical remarks if the topic came up. They knew drawing attention to the issue and creating awareness was not in their interest. Consequently and ironically, climate change coverage dropped even though media awareness increased. However, these trends did not lead to any significant policy action. These historical challenges still haunt US news media climate change coverage, while several new challenges are developing within the system of news reporting that negatively impact public understanding of the severity of the climate crisis. In the larger picture, news media have had at least four to five decades to bring important information about the emergency to the public, and to press for regulation to address the crisis.

Declining Journalists

The restructuring of the news industry after the great economic recession of the late 2000s became a major challenge for journalists writing about climate change. This was mostly because there were fewer journalists able to cover the topic since about 50 percent of journalists were cut around 2010.[20] A few years after that, the number of journalists compared to 25 years ago stood at about one-third, according to Robert McChesney, professor of communication at the University of Illinois.[21] That meant that two-thirds of reporters disappeared from US journalism just in the past few decades. Finding work in the news industry is also challenging as approximately 2,100 newspapers have closed in the United States since 2004, which means almost half of the counties have only one newspaper left, which is usually a weekly. About 171 counties have no newspaper at all, according to Penny Abernathy, the Knight chair at the University of North Carolina at Chapel Hill.[22] For the news organizations still standing, many companies have cut full-time specialists in favor of a workforce of general assignments and freelancers or crowdsourced unpaid amateurs.[23]

An additional round of large cuts to newsrooms and media companies resumed a little over 10 years after the recession and about a year after President Trump gave out his massive tax breaks to the wealthy and corporate America. The theory behind these tax breaks was that they were supposed to help create jobs. By January 2019, HuffPost, BuzzFeed, and Gannett, which owns over 1,000 publications around the country, announced layoffs with 1,400 people losing their jobs. Vice Media also laid off 10 percent of its staff, or 250 employees. In addition, McClatchy, which operates newspapers across the United States, offered voluntary buyouts for 450 workers. In total, as many as 2,100 news and media workers lost their jobs within a few short weeks.[24]

While it may be difficult to fathom the US news media industry getting any worse, it did once again when the Covid-19 virus hit in early 2020. At a time of one of the biggest stories of the century, many US newspapers were losing 30–60 percent of their advertising due to the virus, which induced another lethal round of layoffs and furloughs, leading to even fewer news organizations and less information for citizens. The *New York Times* estimated there were 36,000 employees of news media companies that had been laid off, furloughed, or had their pay reduced after a few months into the crisis.[25] The Poynter Institute kept track of all cuts, layoffs, furloughs, and closures during the initial months of the crisis and, as of May 20, 2020, there were about 150 news organizations on that ever-expanding list.[26]

Gannett, the largest local newspaper chain in the United States, furloughed staff. Alden Global Capital, which owns more than 50 newspapers, issued layoffs, and Lee Enterprises, which owns newspapers in 25 states, also noted it would induce furloughs. The Louisiana-based *Times-Picayune* and *Advocate* furloughed 10 percent of its staff. The Cleveland, Ohio newspaper the *Plain Dealer* laid off 22 newsroom staff, and Seattle's *Stranger* magazine suspended publication while temporarily laying off 18 employees. With only a few weeks into the crisis, industry experts warned the virus would spell the end of hundreds of news organizations, leading to an "extinction-level" event.[27]

This meant that the relatively small number of newsroom workers left from the recession cuts would be cut once again. There are now fewer and fewer people researching and writing for news organizations. For every remaining US journalist, there are at least four times as many people working in public relations, whose job it is to supply the media with content favoring their organizations.[28] Many of these people work within the fossil fuel and consumer industries.

For the journalists and organizations that survived the massive cuts, they now have tighter deadlines and also spend more time conveying information through a number of different platforms/media (like video, audio, and social media).[29] The development of digital media further undermined claims that media structures were naturally supportive for news professionals.[30] These trends mean that there are now fewer and fewer environmental journalists left in the United States to cover climate change. In fact, of the two recent studies that interviewed climate journalists, one study could find 10,[31] while the other found 11.[32] Given these numbers, it appears there are more climate denialist scientists than there are specialized environmental journalists who may be able to check their claims. It is also important to remember that journalists work some of the longest hours for the lowest pay while working in hierarchical institutions where decision-making often happens top-down. They work in complex systems with multiple pressures and stakeholders, where ultimately the bottom line is revenue generation.

Being a journalist with an assignment to cover climate change would be extremely challenging for anyone under these circumstances, especially with the majority of journalists recipients of fewer degrees from science or math than the average US resident.[33] Moreover, many environmental reporters indicate they are not scientists or experts in the field.[34] They often lack significant training in climate science and regularly lack the appropriate amount of time to meaningfully research and evaluate various claims within scientific information.[35] Investigating multiple scientific claims for veracity would be challenging,[36] especially as many journalists do not understand the concept of "peer-reviewed" science.[37] This can create misinformation as well, as one study found a journalist blaming GHG emissions for global warming instead of human activity or elevated consumption.[38] Another study found that the causes and potential solutions of climate change are less likely to be reported than the descriptive terms used to designate climate change. It noted that within stories mentioning climate change, the percentage of stories that contained the words "economy," "cars, vehicle or automobile," "fossil fuel," "energy conservation," "alternative energy," or "renewable energy" were extremely low compared to "global warming," "science," and "Kyoto."[39] In general, growth economics, industrialism, and elevated consumption levels are not tied to producing harmful emissions, while mitigation policies and research that suggests minimizing consumption and regulating businesses are usually ignored by the news media.

These trends create an environment where even the best-intentioned journalists may struggle to create the most thorough climate articles, and some of these struggles appear to predate the post–Great Recession era. Even going back the 1970s accurate portrayals of science in newspaper stories were the exception rather than the rule.[40] In sum, the lack of financial support, time, training, and academic/scientific background is limiting journalists' abilities to cover climate change. Unfortunately, within this new structure, journalistic dependence on information provided by public relations firms is likely to increase.[41]

For those journalists who are able to cover climate change, they can receive an extreme amount of criticism, not only from the public but also from sources and editors. For example, many experience a powerful backlash through flak from a small but vocal conservative right-wing constituency.[42] The criticism can include hate mail, death threats, angry phone calls, among other forms of communication, with the majority of these comments from global warming denialists.[43] The sum of the violence and threats against journalists now has the United States classified as a problematic place for journalists to work, according to the World Press Freedom Index published by Reporters Without Borders. Much of this classification is driven by inflammatory comments made by Trump, combined with a broader hatred of the media, which has influenced violent threats and acts against US journalists and news organizations.[44] Trump has continually called the press un-American and the enemy of the people.

Much of this pressure is difficult to fight, and countering climate misinformation may lead to accusations of being biased, which can lead to demotion or a dismissal. The majority of journalists almost everywhere have been for many years employees of enterprises with bureaucratic hierarchies that allocate tasks, determine rewards, and possess the power to hire and fire.[45] In a field where there are not many jobs left compared to previous decades, losing one's job is an incredible burden. According to Glenn Greenwald, Pulitzer Prize-winning journalist, "One of the many problems with the US media is that as jobs disappear, young journalists are increasingly deterred from questioning or challenging prevailing media narratives because it can be career-destroying. The overwhelming incentive is to affirm media pieties or stay quiet."[46] One has to ask if silence or self-censorship may be preferable especially when journalists have families to feed and bills to pay. Indeed, like most people, environmental journalists also exist within a world where they have their own economic insecurities.[47]

Fortunately, not all environmental reporting exhibits limitations. There are phenomenal examples from environmental journalists in media, and not everyone or everything is a barrier to great journalism about climate change. Many journalists can, have, and continue to create impressive climate change news pieces. The issue is that the phenomenon of climate change is an incredibly complex multidisciplinary topic. The challenges for those who work or research in the field are multifaceted, while journalists perhaps have an even greater burden of responsibility in order to inform the public about important issues. Climate change is difficult because disbelief in human influence, complexity, uncertainty, invisible causes, distant impacts, lack of immediacy, lack of direct gratification for taking mitigative actions, self-interest, and inadequate signals indicating the need for change all complicate climate change coverage.[48]

Decline in Coverage

The recent cuts from the Great Recession influenced the decline in the number of journalists and negatively impacted climate coverage. The industry also consolidated to the point where roughly 80 percent of news comes from only four companies that control 80 percent of the national distribution of news in the United States.[49] While the fossil fuel industry continues to battle against regulation to help mitigate global warming, an overall trend toward declining climate change coverage in US and global news media, particularly since 2010, continues to develop.[50] This overall negative quantitative trend in coverage is concerning and no doubt aids the fossil fuel industry. According to a report that combined 27 leading academic institutions, the UN, and intergovernmental agencies from every continent, overall global newspaper climate change coverage declined at an average rate of 1.25 percent per year between 2007 and 2017.[51]

An additional study found major US newspapers significantly decreased their coverage of climate change from 2007 to 2012, from roughly 2,500 articles in 2007–8 to 993 during 2011–12.[52] This change represents more than a 50 percent reduction in major US newspaper climate articles.[53] According to a study by Media Matters for America, between 2015 and 2016, all of the major US evening news broadcasts and Sunday shows on NBC, CBS, ABC and Fox collectively decreased their total climate change coverage by 66 percent.[54] Things were so bad that ABC's *World News Tonight* did not feature a single scientist in its climate coverage. The decline happened even when there were ample opportunities to discuss it. For example, 2015 was the hottest year on record, there was the Paris Climate Agreement, there was an election on the horizon, and there were several extreme weather events. Nonetheless, the media still found time to uncritically air climate denial, most of which came from President Trump and his administration.[55] During the next few years, the downward spiral in broadcast coverage of climate change continued even further. Between 2017 to 2018 it declined by another 45 percent, even as the climate crisis continued to worsen.[56]

Another study found local coverage of climate change decreasing by 47 percent in the *Pittsburgh Post-Gazette*, and by 78 percent in the *Pittsburgh Tribune-Review* from 2016 to 2017.[57] The same study found a slight increase in climate change coverage for national newspapers. Moreover, a small segment of coverage—health and climate change—also increased by an average of 4 percent per year between 2007 and 2017.[58] What the studies didn't note about the slight increases is that when the Trump administration deregulates environmental policies or hires climate deniers, they unintentionally increase publicity about global warming if news media cover it. In quantitative terms, this content would be positive assuming overall coverage about the climate crisis benefits society. This is ironic given that while the news media's sources are elite driven, when elites act in opposition to policies that positively impact the environment, their news content counts positively. These articles, while educational, do not necessarily promote the immediate collective action through government regulation that is required to address the crisis. Additional reporting on ways in which citizens can be empowered to address the climate crisis combined with editorials supporting immediate regulation and policy change should be encouraged.

In this sense, quantitative analyses of climate change reporting are helpful in that they shed light on the number of articles that mention key terms. In general, it is safe to assume the higher the quantity of articles covering the topic, the more likely the public will be informed about the issue. However, quantitative data can sometimes obfuscate the angle, topic, theme, and framing of articles, so it is important to look at communication and climate change research through a multitude of methodologies and perspectives. If the media publish significant climate information from deniers and constantly report on how many ex-fossil fuel executives the Trump administration hired in his EPA, these articles easily fall under categories about climate change. While they are helpful (except for the denial misinformation), there still needs to be more about the urgency of the crisis and the need for immediate policy changes.

There is also a significant quantitative gap between US coverage compared to countries like the United Kingdom, Canada, and others. Overall US media coverage, and in particular that of the *New York Times*, continues to be behind the coverage of other countries around the world. Comparing mainstream/agenda-setting newspaper coverage of climate change in the United States and the United Kingdom between 1988 and 2006 shows that the number of UK articles was always more than that of the United States. For example, in 2003, during a rare year where climate change coverage actually increased, the number of UK articles skyrocketed from 750 to well over 2,000, but US newspapers only slightly increased their coverage from roughly 500 articles per year to around 750.[59] Clearly there is a quantitative difference between the newspapers of each country, with some in the UK taking climate change more seriously than the United States. Moreover, the coverage gap is not only with the United Kingdom. US newspapers (which includes major newspapers apart from the *New York Times*) are also less than one-third as likely to have global warming or climate change stories compared to Canadian newspapers and about two-thirds as likely to have those stories than international newspapers.[60] That means international newspapers are about 30 percent more likely, while Canadian newspapers are three times as likely, to have climate change stories compared to US newspapers. Moreover, an additional content analysis of the *New York Times* (US), *Le Monde* (France), and the *Frankfurter Allgemeine Zeitung* (Germany) newspapers showed significant differences in coverage between 2007 and 2017. The European newspapers experienced a decline in the proportion of articles explaining and justifying why climate change is a public health issue, while simultaneously experiencing an increase in those highlighting the health dimensions of national climate change interventions. By contrast, in the *New York Times*, the vast majority of articles (92%) referring to both health and climate did not link the topics.[61] Chapter 10 expands on this international comparison by examining the qualitative differences, instead of the quantitative differences, between the US and European press.

False Balance Coverage: False Balance Leads to Bias

Besides the decline in the number of journalists and climate change coverage, the journalistic norm of balance also challenges climate change news by creating false balance, which leads to bias. When comparing coverage to many other industrialized Western countries, US mainstream news media coverage of climate change has one of the worst records largely due to its history of supporting extreme denialist or skeptic minority views. This gap in coverage manifests itself through the manner in which the news media construct the concept of global warming. In essence, the idea that every story needs to be equally balanced, regardless of their scientific merits, created a situation where balance invented imbalance, or false balance. By creating a "balance" between denialists and scientists, the news media greatly distorted public understanding of climate change by enabling fossil fuel industry spokespeople, or their sympathizers, equal coverage in global warming reporting. Here, journalists gave the 1–3 percent of climate scientists who largely deny global warming equal coverage to the 97–99 percent of scientists who unanimously agree that global warming is real and that it is human influenced. This kind of false balance was particularly egregious on TV, where one study found that, between 1996 through 2004, about 70 percent of US television news segments provided "balanced" coverage of anthropogenic climate change.[62] Several of the years leading up to this time frame also showed the news media decreasing the use of scientists as sources for the reporting.[63] These trends were widespread during the 1990s and the 2000s.

Another study that examined news media between 2011 and 2012 in the Great Lakes area found significant climate change deniers paired with supporters. In this case, nonscience sources dominated coverage of climate change reporting. This is troubling as the study found that the vast majority of news sources (82%) that doubt human causes of climate change were from nonscientists.[64] It found that reporters who wrote only one climate change story were most likely to *only* use deniers or supporters in a single story. For those who wrote two-five stories, they were most likely to run stories with only denier sources. Those who wrote more than six stories were most likely to use both denier and supporter sources. Moreover, deniers were used more often than supporters in leads.[65] This is surprising as the most credible sources regarding climate change should consist of scientists themselves.

Denier groups continue to enjoy extensive media exposure today and have significantly increased their presence in multiple mainstream media outlets during the past two decades.[66] Since the Trump administration, the US media have doubled their coverage of these groups compared to their reporting over the previous decade, with the *Washington Post* and the *New York Times* leading the way. Ironically, conservative media like *Fox News* and the *Wall Street Journal* are much less likely to cover climate change countermovement groups. Scholars Maxwell Boykoff and Justin Farrell suggest the "naming and shaming of right-wing outlets for amplifying the volume of CCM organizations and their associated contrarians is misplaced."[67]

Commercial pressures within news media partially explain and contextualize the origins of false balance. One way this presents itself is through the need to garner controversy, which can make for more exciting reading or viewing, which theoretically should increase profits through enlarged readership and viewership. On occasion, journalists under pressure to garner high ratings may develop controversy where none previously existed.[68] Situating speakers in terms of opponents and proponents, similar to a boxing match or some sort of dual, tends to fit these categorizations.

The boxing match approach to "balance" perspectives often frames environmental reporting though personalities, like Al Gore versus Sarah Palin. These professional norms of dramatization, novelty, and personalization also influence deficient coverage.[69] Ironically, following these behaviors can decrease the accuracy of proper, precise, and pressing climate change coverage. Framing environmental issues through personalities may be able to influence news media profits, but it also individualizes responsibility for a challenge that can only adequately be addressed institutionally and downplays any sort of structural analysis. Last, many of these approaches can take the conversation away from the role of power relations, context, and processes and instead replace them with individual characterizations of claim makers.

Besides commercial pressures to create conflict for higher ratings, some journalists or editors may see "balance" as a replacement for an inability to check scientific claims. Checking scientific claims is time-consuming and, for many news corporations, giving journalists plenty of time to research and fact check is an expensive endeavor that can cut into profits. Given the time constraints on journalists, the easiest climate change story for the overworked generalist reporter (which is now the stockholder and media manager preferred identity) is to write the false balance story.[70] Fewer journalists covering climate change, combined with hurried industrial, factory-like speed, and demands from news organizations and managers, means that the public may still see more false-balance coverage, simply because it's faster and produces a story that fits within the professional journalistic norms and appeases ownership and management.

Political Orientations

Apart from time constraints, journalist and media owner political orientations can also influence false balance. A recent study found that the more conservative the journalist, the more likely he or she was to grant climate change skeptics equal treatment, while those who wanted to give equal voice to skeptics used less scientific sources.[71] Fox News, which is the most popular cable TV news network, consistently airs more stories that question anthropogenic causes of climate change than stories that accept these scientific claims.[72] Exposure to conservative media is negatively related to believing in global warming and supporting mitigation policies,[73] whereas the use of nonconservative media is associated with greater acceptance of the science and of the impacts of global warming.[74] Another study found that Fox News influenced climate denialism as it decreased viewers' trust in science and global warming. It accomplished this after presenting inaccurate information. For example, in 2013, only 28 percent of Fox's climate science news segments were accurate. This was, however, a significant improvement over 2012, when it had a 7 percent accuracy rate.[75] However, mainstream media, instead of conservative media like Fox, tend to give deniers more publicity according to another recent study.[76]

Right-leaning news organizations also often embrace alarmist discourse, but only in order to undermine government action with skeptical constructions by drawing on climate change denial or comic nihilist repertoires. The "liberal" press may also use alarmist rhetoric, but then mount a call to some sort of personal or small corporate action to address the crisis.[77] Some of the right-wing media also convey climate coverage through free market protection repertoires with conjecture that warming will be good, instead of calling for institutional or government policies. Newspapers, in particular newspapers on the right of the political spectrum, can regard mitigation plans as a threat to consumerism and market preeminence.[78] Leaving out information, creating misinformation, or downplaying these voices, which is perfectly legal in a deregulated media environment, also influences the false balance approach.

Last, when politicians discuss climate change, journalists are more inclined to view their story as a political story instead of a climate change story and implement their historic false balance coverage. Reporters throw out the "weight-of-evidence" and the two "sides are equal" approach comes back into play. Here, skeptics and denialists are given plenty of news media space and time because they are politicians who "question" the human influence of global warming. They are not denialist scientists, they are denialist politicians. They may use the same talking points given to them by the fossil fuel industry think tanks and public relations firms, but for US journalistic norms, the topic can revert to a political story, where the denialist perspective is given a prominent voice in the news media. Now that many journalists have slowly adapted and largely refuted denialist claims, this type of story allows the fossil fuel industry to implement a strategy of using politicians as their spokespeople if they are forced to talk about global warming.

Having politicians as major contributors to the official sources bias of US news media helps this process. Moreover, journalists may self-question correcting politicians who create climate misinformation because reporters can be seen as biased against whatever political party is making the claims. While denialist politicians often get a green light from the media, climate activists, many of whom are trying to expose misinformation, "are to be approached with caution" according to some journalists.[79] This is not necessarily a bad thing as the fossil fuel industry sets up front groups though paid public relations firms, which would need to be scrutinized. Climate activists may construct their own interpretations of science, but either way, the news media should always scrutinize sources, including activists or politicians. In general, the media are most likely to negatively cover activist groups, while simultaneously they are more inclined to include corporate/

industry and government sources in their reporting as opposed to citizen or activist sources. The official source bias harms mediated discussion of important topics.

News Media Silence: Editors and Advertisers

In addition to the declining number of journalists, decrease in climate change coverage, and false balance, editors and advertising interests can also challenge overall climate reporting. With most news media profit-oriented and dependent on advertisers for funding, editors and advertising interests often clash with many of the topics that arise within climate change reporting. Many journalists around the world believe mainstream media editors do not prioritize coverage of climate change and instead focus on more sensational stories evolving around crime, violence, and other topics that garner higher ratings and approval with advertisers. They also note editors sometimes downplay topics that could compromise advertiser interests and that resources for conveying climate change information are scarce.[80] For an analogy about how much influence advertisers can have in the news media, well-known muckraker journalist and editor Charles Russell in the late 1800s noted that as commercialism and consumerism evolved, the newspaper quickly became an appendage of the department store.[81]

Indeed, reporters point to the challenge of getting editors on board with climate change coverage, as some editors believe audiences struggle with the subject, which means editors may be less interested in pursuing the topic.[82] This trend with editors challenging climate change coverage is global, as another study that interviewed journalists in four developing countries found that media organizations have a poor understanding of the so-called climate change "debate" and that many express little interest in it.[83] Climate change stories can also consist of depressing, bad news, which are less likely to bring in higher ratings that advertisers favor. According to MSNBC's Chris Hayes, when news media cover climate change it can kill their ratings.[84] Indeed, the bias of market-based media is that the market causality of the media agenda frequently excludes topics that don't sell.[85] So when there is consensus and the news is depressing, the topic may not fit within commercial, profit-oriented news media systems.

Having a predominately commercialized advertising-supported news media, like in the United States, means that the system is mostly funded by the consumption of products and services that are advertised through the media. Why would a business that profits from consumption want to pay for advertising in a newspaper or TV news channel that often points out the negative environmental repercussions of a consumer-oriented growth economy? It would not make sense. For example, Procter and Gamble, which has consistently ranked as the number one or two advertiser, has had a long history of using its influence over media content when it questions its consumer-oriented goals. Going back to the 1930s, Procter and Gamble canceled its advertising in newspapers when a columnist explained how to make soap at home, which would cut down on consumer spending. Procter and Gamble also developed a policy in the 1950s making sure it would not sponsor media content that depicted industrialists in a bad light. This policy continues through today.[86] Moreover, US car and fuel companies threatened to withdraw advertising from various radio programs after they reported on climate change.[87] They also threatened a TV editor to withdraw car and oil advertising following a report connecting a large flood with climate change.[88] Indeed, a "dependency on advertising revenue from fossil fuel industries may lead journalists to self-censor their stories so that they are written in ways that do not directly threaten their interests," according to scholar Alison Anderson.[89] As the old saying goes, don't harm the hand that feeds you.

To focus even a few stories on the environmental implications of consumer-oriented growth economics could mean a demotion or a chance to find a new job somewhere else as this topic directly questions the businesses that pay for advertising in the media. With consumption and economic growth influencing the climate crisis, these issues are often downplayed in the flagship *New York Times*, particularly regarding energy and nuclear energy.[90] The likelihood of these issues getting decent coverage in the US press is possible, but pretty close to nil. According to Maxwell Boykoff, "with such socio-political concerns at the fore, greater stress is placed on the danger of climate policy on trade and the economy, rather than also considering how trade and the economy may have detrimental effects on the global climate."[91]

Why would a commercial news organization want to devote significant resources to a topic that could cause a decline in revenue to its own company? It is important to remember that media corporations are like other corporations in that they are profit-oriented and have to please investors and board members. Reporting on climate change or taking an editorial stance for divestment of fossil fuel can mean fewer profits for corporate media owners. For example, *USA Today*, along with other media outlets, are owned by Gannett, which is the largest newspaper publisher in the country as measured by total daily circulation. Gannett recently merged with GateHouse Media, owned by Wes Edens's New Fortress Investment Group, which also owns New Fortress Energy. New Fortress Energy deals in natural gas and invests in fracking. Fracking releases methane gas, which is incredibly more potent as a heat-trapping gas compared to carbon dioxide, and thus negatively contributes to global climate disaster.[92] When media corporations are directly or indirectly tied in to the outcome of profits from the fossil fuel industry, and science tells us that it's time to divest from this industry, a natural conflict of interest develops that can directly or indirectly affect editorial decisions. Moreover, threats of financial losses due to advertiser exits from energy corporations can also cause significant challenges to these goals. One can see the commercial nature of the press as a barrier, the funding system as a barrier, the hierarchy as a barrier, and the ownership structure as a barrier to creating quality climate change reporting.

Nonetheless, within many limitations, some benefits of this system do surface. With more and more income coming into news media from subscriptions—for example, subscriptions from the *New York Times* account for roughly two-thirds of its income—dependency on advertising and consumerism may diminish. Being less reliant on advertising is a good thing for journalism, and it could open new avenues for climate change reporting. Even though public access to information is challenged with subscriptions, they could create an environment that encouraged more articles regarding consumerism and its relationship with the climate emergency. However, over 100 years of advertising dependency doesn't disappear overnight and even with only one-third of the budget of the *New York Times* coming from advertising, it doesn't mean acquiring more isn't a priority. The *New York Times* did not appear to be improving its climate change coverage even as it transitioned toward higher income from subscriptions. Climate activists, who organized a massive protest in 2019 in front of their New York office building, were still not satisfied with the paper's coverage.

Even with subscriptions, the news media still have a ton of advertisements that promote a consumer economy. The quantity of pro-consumption messages in commercialized media systems vastly outguns any messages that question the relationship between elevated levels of consumption and environmental degradation. To compare this battle between messages is like having one side armed with a nuclear weapon while the other side has a baseball bat. It is clearly an asymmetric information battle with disproportionate approaches. The media may have an article or a news segment for a few seconds on climate change, but then the next page will promote a sports utility vehicle, while the following commercial will promote the Daytona 500. Additional advertisements will most likely consist of the latest and largest flat screen TV and of vacation getaways via an airline. All of these are GHG-intense items and experiences.

Nonetheless, many reporters have been very creative and successful in terms of resisting pressure from editors against climate change stories.[93] Even with these challenges, they have been able to get climate change stories out to the public, often through weaving climate change into a diverse spectrum of beats and specialties. However, there is some concern that this approach may dissolve the central focus, immediacy, and crisis of climate change.[94]

News Media Silence: Silence Reporting

The sum of the trends discussed in this chapter so far no doubt influence climate silence reporting and overall news media silence regarding climate change. Climate silence happens when news media ignore climate change, especially by deciding not to connect extreme weather or other environmental phenomena with global warming. This can be regarded as a form of climate denial and it appears to be a relatively recent trend, as many previous studies concluded that environmental coverage is most likely to be event-driven. Without any concrete incident that can be directly connected to an environmental issue, it would be less likely for the media to report on it.[95] However, the event-driven coverage seems to be decreasing as it disconnects these events with global warming. Robert Hackett, professor in the School of Communication at Simon Fraser University, confirms this media silence when he found that alternative media and environmental communication practitioners find the decreasing importance of climate change within mainstream media a problem that is increasing.[96] They also find that much reporting does not connect the dots between events, weather, and policy among other influences. In addition, when one examines extreme weather events combined with decreasing or different energy use, the frequency of the stories dramatically decreases.[97]

To better demonstrate climate silence, one study from the progressive media watchdog group Fairness and Accuracy in Reporting, or FAIR, found that during a massive heat wave in December 2015, where close to 700 warm records were set through much of the Midwest and East Coast, very few media outlets linked it to climate change. Of 259 newspaper stories that covered the heat wave between December 1 and December 14, only seven linked it to climate change. Television network news largely blocked out connecting climate change to the heat wave. Out of 67 mentions, only one talked about the climate connection.[98] An additional report by Media Matters that tracked coverage of an extreme two-week global heatwave during July 2018 found that out of the 127 mentions of the heatwave by corporate broadcasters ABC, CBS and NBC, climate change was mentioned only once.[99] During the same time period, the Public Broadcasting System, or PBS, did air two segments on the heatwaves and one did discuss climate change.

Climate silence doesn't stop with heatwaves. Hurricane Matthew, which hammered the US East Coast in October 2016, illustrates another example of media leaving out climate change connections. This happened again two years later in 2018, when Hurricane Florence came smashing across the United States, triggering evacuation orders for over a million people. The news media again failed to discuss one of the most important underlying influences regarding the strength of the hurricane. According to a study by the Public Citizen, 7.5 percent of Hurricane Florence stories in the 50 top-circulation US newspapers between September 9 and September 16 mentioned climate change with only 4.3 percent of major TV outlets mentioning it. While Fox News did mention climate change, its segments denied the relationship between Florence and climate change.[100]

Moving forward to 2019, with nonstop media coverage, Hurricane Dorian destroyed significant parts of the Bahamas but again, like previous years, few if any media linked the hurricane to global warming. ABC, CBS, NBC, CNN, MSNBC, and Fox connected the hurricane to climate change only 12 times in 167 pieces.[101] Another study found that broadcast news aired just one piece, out of 216 segments from August 28 to September 5, that mentioned links between climate change and hurricanes like Dorian.[102] Moreover, of the 363 articles about Dorian in the most circulated 49 newspapers, only 9 mentioned climate change.[103] With hurricanes becoming more dangerous for coastal areas through increased ocean temperatures and higher sea levels—direct results of global warming—very few media outlets mention global warming in response to hurricanes. They also fail to mention how human involvement has influenced climate change.

This is ironic as one would assume climate change would be front and center in media discussions of hurricanes, especially during an election year. However, this did not appear to be the case. According to May Boeve, executive director of 350 Action, the political arm of the climate organization 350.org, "we're seeing a deafening climate silence from journalists and in this presidential cycle, where there have been no questions asked in the debates about this issue."[104] Climate silence then continued after the elections swept into power the world's best known climate denier. FAIR found that for several months after US President Trump took office on January 20, 2017, not one media outlet asked the president a question about climate change during their interviews with him. In stories from 2017, AP News (April 21), CBS News (April 30), the *New York Times* (April 5), *The Economist* (May 11), NBC News (May 11), ABC News (January 25), Bloomberg News (May 1), Fox News (February 5), Breitbart (February 27), Reuters (February 24), *Time Magazine* (March 27), and the *Financial Times* (April 2) all failed to ask Trump about his climate change views or policies.[105]

The hiring of climate skeptics by agenda-setting news media does not appear to be correcting the silence. For example, in 2017, the *New York Times* and NBC News hired climate denialist Bret Stephens, who is also an MSNBC contributor. He previously worked at the *Wall Street Journal* and the *Jerusalem Post*. These media corporations have given him a prominent voice on the subject while simultaneously helping confuse and miseducate readers and viewers about global warming.

Importance of Supplying Studies

Climate silence does not mean that all corporate news media outlets censor climate change. Publishing information about well-known studies that are supplied to the news media occurs periodically each year. The studies that are published appear mostly limited to large or well-known governmental or international agency studies like the IPCC, which are fed to the media usually only once a year. However, when these studies are submitted to the news media, many news organizations, if not most, still do not cover them. For example, when the IPCC released its 2018 landmark report that noted humanity had only 12 years to radically diminish GHG emissions or face a questionable future on the planet, only 22 of the 50 biggest newspapers covered the report.[106]

Nonetheless, these studies do receive some coverage within the constraints of a corporate, profit-oriented news system for a few reasons. Dependency on large global warming studies helps the media not only with free content but also with a justification to publish articles about the topic in the first place. The media can publish these studies due to their stature and news value, and they can deflect responsibility and critique back to the group that created the study in the first place—and away from themselves to a certain extent. This helps when climate deniers attack the media. Ultimately, with the corporate media largely based on supply rather than demand, larger national and international climate science studies may want to increase the number of studies they send the media, or cut their larger studies into several smaller ones and release them throughout the year in order to increase media exposure about the global warming crisis. The frequency of the rate in which these studies are released no doubt impacts media exposure. The Trump administration knows this and as a result has implemented measures to decrease the release of US government climate change studies.

While sending additional studies more frequently to the news media will help, climate silencing also goes beyond the media. It is also occurring within political debate and particularly from Republicans. Silence as a political strategy no doubt influences media silence. For example, after the Republicans won enough seats to control Congress during the Obama administration, one of their countermovement goals was to change tactics by minimizing official discussion of climate change in hopes of keeping it out of news media. When one of the two political parties uses silence as tactic, the media tend to follow. In other words, as the Republican political strategy of climate silence not only affects policy toward mitigating global warming negatively, it also negatively affects information and public debate as the media are most likely to follow through with media silence due to the sourcing of their stories. This relationship also exists in the United Kingdom, as one study found that political actors seemed to play a more powerful and effective role in shaping climate change in the so-called public sphere during the 1980s through 2003.[107]

There is also not enough attention to responses or solutions to the climate crisis and instead there is too much focus on political competition or conflict among elites if and when climate change is treated as a political issue. Of course, there is little focus regarding who or what is responsible for climate change, nor is there much analysis of capitalism or critique of social structures.[108] Instead, news media, like in the United Kingdom and Australia, have a tendency to focus on insurance-related aspects during catastrophic weather-related events while downplaying the connections of broader climate patterns with global warming and climate change.[109]

Moreover, news media climate silence also appears to be limiting interpersonal discussion about climate change. Many Americans say they don't hear about it in the media, from people they know, or from family and friends. For most Americans, climate change is not a common topic of conversation, or something they hear about much in their daily lives.[110] In this sense, many Americans, even those who care about the issue, shy away from discussing it because they rarely hear others talking about it. This trend has led some researchers to believe there is a climate change "spiral of silence," which decreases the amount of dialogue.[111]

Climate silencing, where the media either black out or fail to connect weather events with global warming, is problematic as the public has a right to know what is happening around them and if these events are harmful. The public does need to know, and if political policy solutions may be able to mitigate some of the effects of global warming, then media silence on critical issues is a threat to democracy, as the right of citizens to be informed is a crucial democratic tenet.

The next chapter continues examining the relationship between journalism and climate change. It looks at the powerful fossil fuel industry's misinformation initiatives, and at blogs and their relationship with public opinion, noting several factors that influence public salience of climate change misinformation. Nonetheless, as the science of global warming becomes more forceful and dire, a trend toward more public understanding and support of initiatives to address global warming is developing.

CHAPTER NINE

Misinformation, Blogs, and Public Opinion

While the previous chapter focused on the declining number of journalists, decrease in climate change coverage, false balance, and the circumstances that influence news media silence with climate change reporting, this chapter continues to outline influences that constrain the quantity and quality of climate change journalism. I begin by discussing the relationship between the fossil fuel lobby, media misinformation initiatives, and the lobby's success in delaying government regulation. Next, I discuss several reasons why climate skeptics are inclined to believe the fossil fuel lobby's misinformation, or at least question the mainstream news media regarding global warming. Commercialization of the media, the myth of the liberal media, politicization of climate science, the cocooning of the internet, as well as problems within political and economic systems all influence public salience of climate change (mis)information. Last, these trends indicate a decline in public concern about global warming, with some recent studies suggesting this trend may be flipping, as a significant percentage of Republicans now appear to believe global warming is happening.

Fossil Fuel Lobby and Media Misinformation Initiatives

Conservative think tanks and pressure groups, such as the American Enterprise Institute, the Heartland Institute, the Heritage Foundation, the Committee for a Constructive Tomorrow, the Center for the Study of Carbon Dioxide and Global Change, the European Institute of Climate and Energy, Cooler Heads Coalition, Global Climate Coalition, Science and Environmental Project, the George C. Marshall Institute, the Competitive Enterprise Institute, the American Legislative Exchange Council, and the Science and Public Policy Institute, among several others, are some of the main contributors to climate science denialism and misinformation around the world. The denial industry has also worked closely with the US Chamber of Commerce to oppose bills that would begin to address GHG emissions.[1] These groups are just a small slice of the overall number of organizations that form a climate change countermovement. A recent study found there are 4,556 individuals tied to 164 organizations in the United States that spread misinformation and disinformation about climate change.[2] ExxonMobil and the Koch Family Foundation have long been prominent backers of these groups.

These think tanks and organizations are part of the polluter-industrial complex, which is a complex infrastructure led by the most powerful corporate polluters in the United States.[3] Besides think tanks, other policy institutes, research centers, legal associations, public relations firms, political action committees, and other entities all work to discredit the environmental movement and dismantle state programs and policies that safeguard public health and the environment.[4] The polluter-industrial complex is highly connected to the state, profits from neoliberal environmental policies, and benefits from its campaign funding, lobbying endeavors, and efforts to regulate scientific standards.

Their output, along with attacks against science, increased from 1998 to 2013.[5] Many of these groups and others received around $125 million during a recent three-year time span by secretive donors.[6] Secretly donating through the Donors Trust and Donors Capital Fund helps these anonymous groups avoid bad publicity. They work to defeat climate bills in the US Congress, mobilize against EPA rules that can reduce carbon pollution, work to roll back measures that promote wind and solar power, and block measures to plan for future sea-level rise.[7]

Much of the misinformation about climate change generated from these groups is largely attributed to the power and money of the fossil fuel industry. The massive power and influence of the fossil fuel industry's public relations arm no doubt grabs the attention of the news media. Campaigns against the science of global warming from corporate think tanks were generally viewed as credible and accepted by most journalistic practices during the past several decades, which in turn have been effective in terms of influencing public opinion.

With the news largely based on supply, not demand, it is easier for the fossil fuel industry to provide news and information with their own data, spokespeople, and studies. The practice of using official and authority-order sources, which leads journalists to consult authority figures such as government officials, business leaders, and think tank spokespeople at the expense of non-elites, challenges journalists' abilities to accurately report on climate change and boosts fossil fuel media relations. Journalists tend to listen and take their input seriously. Indeed, most sources in previous climate change reporting consisted of business, government, and interest-group sources.[8] When a large corporate client sends its spokespeople to "balance out" the mediated global warming debate, standard modes of operation make this perfectly reasonable. Moreover, most of the largest corporations in the world promote their "corporate responsibility" initiatives, many of them environmental, where they praise their own activities. These corporations often appropriate environmental and social improvement discourses for public relations purposes.[9] The success of these corporate public relations initiatives has had a tremendous influence on public misunderstanding of climate change.[10]

The fossil fuel denial industry funds several misinformation strategies to confuse the public, while simultaneously influencing opposition to government regulation of industries that can help mitigate the climate crisis. ExxonMobil, Shell, Chevron, BP, and Total spend about $195 million a year on branding campaigns that suggest they support action against climate change while simultaneously spending massive sums of money to prevent action against global warming. According to a report by InfluenceMap, the five largest publicly traded oil and gas corporations (ExxonMobil, Royal Dutch Shell, Chevron, BP, and Total) invested over $1 billion of shareholder funds during the three years following the Paris Agreement on misleading climate-related branding and lobbying.[11] Much of this has been spent on influencing media and social media, as well as on opposition to climate related ballot initiatives. The billionaire Koch brothers also pour massive amounts of money into climate denier think tanks and politicians, which has also had a significant impact on denialism in the United States. They have one of the largest corporate lobbying entities in the United States and are located just about a block away from the White House. Their main aim is to counter any effort to regulate GHG emissions because their businesses are deeply involved in the fossil fuel industry.[12]

Some of the denial money coming from these groups funds scientists, research, politicians, and organizations to question global warming. A small number of industry-supported scientists and their studies are often used as tools to obscure the truth and create doubt about global warming.

These people and studies are then given substantial media time and cited by politicians, which in turn helps confuse the public and delays action.[13] Indeed, becoming a denier or skeptic can be lucrative for scientists, politicians, and many organizations. For example, a lobby group funded by ExxonMobil offered scientists and economists $10,000 plus expenses for articles that would refute findings from the IPCC in 2007.[14] Many companies that profit from fossil fuel consumption also provide financial support to their political allies in efforts to undermine any sort of public trust in climate science.[15] In the mid-2000s Exxon Mobil also distributed close to $4 million to 39 organizations challenging climate change science.[16]

This money generated different media, which helped create skeptical hype. An analysis of English language environmental books with skeptical orientations published from the seventies until the mid-2000s found that 92 percent of these books were connected to mostly US-based conservative think tanks. These organizations supported skeptical and denialist outlooks concerning environmental science.[17] Interestingly, an additional study found that skeptical films tend to have a greater influence on viewers' opinions than films that try to increase awareness about the threats of climate change.[18] Moreover, knowledge generated from television viewing is reported as weaker than print usage with self-professed levels of environmental knowledge.[19]

One of the main goals of climate denier think tanks is to confuse the public by presenting the illusion that there is no scientific consensus on climate change. Suppressing the idea that there is a consensus, along with questioning beliefs in science, have been strategic mainstays of many groups. Additional strategies, such as exaggerating the costs of mitigation, suggesting that action addressing climate change will harm businesses and the economy, and presenting the claim that only a small liberal elite will benefit from addressing climate change, have also been consistently used. Systematically producing and repeating misinformation with the goal of encouraging misunderstanding has been fairly effective, according to poll data from the past few decades. These polls are examined later in this chapter.

Indeed, the climate denial industry uses several persuasive messaging strategies. An excellent study found that the industry creates four prominent narratives in their efforts to persuade citizens away from government regulation of fossil fuel industries. Their first approach is to frame America's everyman/everywoman as an industry supporter. Their second approach is to frame the president, the EPA, and politicians as uninformed/

unenlightened. Their third is to paint the industry as a paternal caretaker for American citizens, and last is to identify industry supporters as the moral majority.[20] These narratives distract from the negative effects of climate change through the appropriation of US cultural mores and values with the goal of moving public deliberation away from climate change and potential regulatory policies to how these policies would hurt industries, the economy, and supposedly adversely affect average US citizens.

Moreover, the argument that eliminating fossil fuels would hurt the country economically is also well thought out from a propagandist point of view. This argument is even more effective during the current neoliberal period of economic stagnation because more and more people are struggling economically during the past 30–40 years. However, it also points out the hypocrisy of their double standards. The fossil fuel industries get to have the benefits of neoliberal deregulatory economic policies, and that includes all of their subsidies, which also reward the wealthy and corporate sectors, while these policies also move jobs abroad while significantly cutting wages and household income after adjusting for inflation. This is socialism for the fossil fuel industry with rugged and harsh capitalist individualism for workers and citizens.

The industry-led solution is more of the same problem. Not regulating industry, or always regulating them in their favor, will continue these problems and many more. Neoliberal economic policies cause stagnation due to the consolidation of industries and lack of competition. Politicians can then use the stagnation generated from these policies to argue against the concept of government regulation. However, government policies can actually help lower and middle class people, create jobs, increase income, and help regulate the environment. FDR's New Deal is the obvious example, while the proposed Green New Deal legislation also illustrates the positive outcomes that can emanate from government initiatives.

In effect, the fossil fuel industry's strategy has been to co-opt the concept of harm and reverse it to imply change or regulation would be damaging to the country. Among many conservatives and within the fossil fuel industry, potential governmental regulations, which are regarded as the most effective way to address the climate emergency, are viewed as more of a threat to peace and prosperity than the negative effects on the planet and human race from global warming. This means the solution to the problem is worse than the outcome of the problem in the eyes of the industry and many conservatives.

The goal of this extreme and harmful libertarian form of marketplace advocacy is to reassure the public that radical alternatives and interventions are not necessary to address the climate crisis and that if environmental policies were to be instituted, they would have negative consequences. In essence, this is a libertarian fear appeal, with the strategy to scare people into believing the status quo is acceptable and less worse than any form of government regulation. While environmental groups will need to develop more effective countercampaigns, the news media must also do a better job detecting, dissecting, and exposing these subtle, yet persuasive, messaging strategies.

Success of the Fossil Fuel Lobby in Delaying Government Regulation

Much of the misinformation about climate change has had a detrimental effect on climate media coverage and has influenced the delay of any substantial policy to adequately contain the crisis. It has created the impression that the majority of skeptics' talking points are valid, and that their perspective is representative of a large number of scientists, even though that is clearly not the case. With noncongruent reporting good at confusing people and disputing formal consensuses, why should a reader get involved or support any initiative or policy until there is mostly unanimous support? These outlier views contribute to ongoing misleading and counterproductive debates that poorly service the public.[21] Indeed, the media can influence citizens' perceptions of their potential political agency, which can translate into action or inaction,[22] or they can follow through with media silence.

How the sum of this kind of media coverage, and lack of coverage, translates into citizen engagement with climate change should be seriously questioned. While there is no doubt that the news media have at times kept the issue of climate change alive, at the same time one could also note that it has limited the extent to which real change in the organization of society can take place.[23] According to Sir Watson, previous IPCC head, "the way climate change has been communicated to the public has also contributed to delayed action."[24] The private and commercial structure of news industries, particularly in the United States, combined with the massive money from industries and the public relations campaigns that spout misinformation and confusion, has effectively delayed climate change action. In short, the fossil fuel industry was successful in manipulating the news media to promote their propaganda while creating confusion among much of the United States, European, and global public.

Partisan Issue?

In addition to these misinformation strategies, the fossil fuel industry and the Republican Party have also been successful in framing climate change science as a partisan and political issue. For example, Trump embodied this strategy well in an interview with Lesley Stahl of the US news show *60 Minutes*: "We have scientists that disagree with [human-caused global warming] . . . You'd have to show me the [mainstream] scientists because they have a very big political agenda."[25] He clearly politicizes the issues by suggesting "we," as in Republicans have scientists and they disagree with the other scientists, who we are supposed to believe are Democrats. Therefore, science is political and not objective. As a result, all science is debatable.

Within this framing, those in the media that discuss climate change are to be perceived as partisan, which means they must be affiliated with the interests of the Democrat Party if they are supportive of the science. "Objective media" would then need to tread carefully within this framework, and either supply the mythical and largely made-up "other side" to the science of global warming, or simply downplay the topic to be safe out of fear of losing their perceived impartiality. Once climate change has been politicized, then the corporate media become less interested in discussing it.

This is for fear of losing viewers and readers who contribute to their ratings, which then translates into the amount of money they can receive through advertising. There is a possibility that Republicans or others who are climate skeptics, out of a feeling of media bias, would stop viewing or reading the media outlet, which again means financial loss to the corporate media. Years of attacks on many corporate media outlets also play a role in how news is created.

Public Questioning of Climate Science

There are additional reasons why much of the public does not believe, or questions the news media's interpretation of global warming science. First, the commercial nature of the media is a turn off to many because it sensationalizes too many topics for ratings. Second, belief in the "myth of the liberal media" creates the impression that global warming is some sort of partisan attack on conservative interests. Third, the individualized structure of the internet can cocoon people off from different viewpoints, and last, given the number of problems facing the US and other global political and economic systems, following the news about and adapting to global warming can be a luxury for many.

The market nature of the media appears to negatively affect the reception of climate change coverage as media commercialism and distrust of the mainstream media are some influences that push people to be skeptical of media claims. One study found media sensationalism, alarmism, and potential exaggeration played significant factors in cultivating this perspective.[26] As noted before, ratings and competition often dictate or influence sensational coverage and help perpetuate the declining public trust in the media. Climate change gets thrown into coverage of celebrities, Donald Trump's ex-wives, and the latest mall sale. The commercial macro impacts the micro, which in this case is climate change coverage. These market-oriented influences are not likely to help convert denialists to become supporters of mitigation regulation anytime soon.

Moreover, because denialists are predominantly conservative, they may believe the "myth of the liberal media," that the media are supposedly biased toward liberals and their perspectives.[27] This belief would logically conjure the assumption that the topic of global warming is a partisan attack on business interests or conservative values. Therefore, within this belief system, because the sources of the information come from supposed liberals, then their content must be biased toward alleged liberal interests. Unfortunately, for those that follow this belief, they are wrong. The majority of research on media politics suggests the opposite, that the mainstream media are largely supportive of the status quo, and that many editorial endorsements have historically been supportive of conservative candidates and conservative perspectives. Source analyses demonstrate that the media most often include conservative sources over liberal ones. There are very few "liberals" who believe the politics of the mainstream media are predominantly liberal. To be upfront, some reporters may be liberals, but the decision makers—the owners and the editors of most new organizations—are generally more conservative. Some "liberal" social issues may be more accepted today in the media, like gay marriage, but most economic and foreign policy issues continue to be conservative or neoconservative. Mainstream media corporations are also run by economic conservatives who exist within massive profit-seeking companies. They are far from the old stereotype of progressive-voting vegan hippies who drive Volkswagen busses back and forth to work. Moreover, the National Association of Broadcasters, the corporate media's lobbying arm, has also been very supportive of Republicans because they are generally more friendly to their business interests.

Blogs, Tweets, and the Cocooning of Perspectives

With this and the past chapter discussing multiple challenges journalists face within news organizations, perhaps online blogs and social media would be able to pick up the slack? According to a weekly content analysis conducted in 2009, the Pew Research Center found that topics involving global warming earned a much greater share through social media, such as internet blogs and Twitter, than through traditional media (television, newspapers, and radio).[28] This complements previously noted research that found a decline in traditional media coverage of global warming, and it appears that social media and blogs seem to be filling that gap. Social media, blogs, and other smaller online media outlets can assist in expanding public conversations about important issues where the mainstream news media fall short. Along those lines, global warming was one of the top five blog stories ten times since January 2009, but once again did not garner nearly as much attention within traditional media during the same time frame. Social media are beginning to show that the official gatekeepers of traditional news media do not always define important topics and that an alternative is possible.

However, it would be inaccurate to suggest social media and blogs are making up for the limitations of mainstream journalistic practices regarding climate change. Wealthy conservatives, right-wing think tanks, and fossil fuel companies are spending significant resources online in the form of advertisements, blogs, and tweets to discredit global warming science in hopes of preventing action. For example, in the run-up to the 2018 US midterm elections, global oil giants spent $2 million through Facebook and Instagram on targeted advertisements promoting the so-called benefits of increased fossil fuel production.[29] A study from Brown University found that roughly 25 percent of all social media tweets about the climate crisis are generated by bots that distort online discourse by disparaging climate activists in addition to including significant climate science denialism. These bots were responsible for 38 percent of all tweets about "fake science" and many have a large number of followers on Twitter.[30]

Internet information and misinformation can be limited and unhelpful while the web can just as easily mobilize activism against some policies designed to mitigate global warming as it would be to mobilize for those policies.[31] In addition, smaller online outlets often lack resources and conglomerate ownership structures that enable them to more easily promote themselves on- and offline. They also lack capital for investigative journalism and are often unable to defend themselves from threats or lawsuits.

Even with these limitations, as more and more people bail on trusting the mainstream media, some appear to be turning to online blogs, many to most of which contain skeptical orientations. In fact, the skeptic blog community seems to be mushrooming. The blogosphere has been a critical media platform for denialists and skeptics in their attempts to influence public opinion.[32] For example, the website DefyCCC, where CCC stands for "climate change cult," is a climate denier site, whose visitors dramatically increased by almost 2,000 percent in 2018 after they bought the "climate change" search term on Google. Their website also contains significant content from WattsUpWithThat, which is another well-known climate denial site. WattsUpWithThat has received funding from the Heartland Institute, which is backed by the billionaire Koch brothers.[33]

One study found that when people were first looking into questions about climate change, they began viewing blogs promoted by climate scientists, but were put off by their style and turned to skeptical blogs instead.[34] Skeptics' strategic use of the internet has surprisingly earned them several online awards. In 2013 for example, skeptics conducted a successful campaign to nominate WattsUpWithThat for the yearly Best Blog Award, which ended up winning the "Best Weblog of the Year" prize. They also won the "Best Science Blog" for the third time in a row during the same period. Another study of the network of climate change blogs found that the skeptical blog community is actually the largest group.[35] This blog community also publishes the most about topics related to the science of climate change, while "climate change politics" and "US politics" are also disproportionately prevalent in the skeptical blog community compared to other online climate change blog groups. According to these researchers, it appears as if the skeptics are winning the online information war in their efforts to question climate science.

The structure of the internet is also exacerbating the tendency for people to follow their own individual interests online, where they can easily become cocooned. Chapter 6 discussed this in detail. The individualization of the internet is particularly troubling as it can fence people off from different viewpoints from which they can potentially learn from or come to understand why they exist. Advertisers' commercial incentives are driving this trend as an individually structured internet more easily allows them to target potential consumers. For people who question climate change, Google's algorithms will use search histories and other data to designate content and ads, as well as send links about scientists blasting climate change and ads about global warming hoaxes.[36] This is false information designed to sway public opinion against acting on climate change. The individualization of the internet, combined with the lack of confidence in news media, perpetuates the fragmenting of the population into all sorts of groups that are less likely to have much contact with each other. This structure encourages a lack of empathy for different viewpoints. This problem is particularly enhanced with skeptics. Studies of media have found that discussions between accepters and skeptics are often characterized by mutual labeling and unfriendly language, with most of the latter emanating from skeptics.[37]

Nonetheless, skeptics deserve credit for some of their media analyses. In their ability to question mainstream commercial media sensationalism and potential political orientations, the denialists/skeptics are exhibiting critical thinking skills and should be commended. This is a positive trait. However, they should also use these critical thinking skills to assess the misinformation created by the various industry-funded think tanks. In essence, critical thinking is about understanding the power relations behind whatever topic is at hand. With climate change information, it should hopefully become clear that the industry is creating misinformation in order to delay or kill any form of government regulation that could limit their profits or force them to change the status quo. The industry actions come at the expense of the environment and propel the planet toward the tipping point of a no-return warming scenario.

Deteriorating Democracies

Besides the commercial nature of the media, the myth of the liberal media, and the individualized structure of the internet, there is another reason why polling data places climate change near the bottom of priorities for most Americans: the deteriorating political and economic environment of the country. While these polls do not mean people are antiscience or anti-mitigation regulation, they suggest people are facing more dire and immediate challenges in their lives. They are facing pressing problems, like finding a decent job, earning enough money in order to eat and pay rent, and so forth. Also, with oil spills, fracking disasters, floods, wars, crime, polluted drinking water, poverty, police oppression, the murder of gay and trans peoples, institutional racism and sexism, almost weekly gun massacres, terrorism (state and nonstate), lack of access to affordable health care and decent education, and exposure to various toxins, US citizens do not have the same regulatory systems and benefits that many countries in Europe and some in the Global South enjoy.

In some ways, the ability to be very concerned about climate change in the United States may be a privilege given the levels of unemployment and poverty for many. Economic trends no doubt contribute to the overall public perspective of the threat imposed by climate change.[38] According to a recent analysis of US government data by the nonprofit United Way Alice Project, 50.8 million US households, or 43 percent of households, can't afford a basic monthly budget for child care, health care, food, housing, transportation, and a monthly smartphone bill.[39] Besides these day-to-day challenges that most likely downplay the significance of global warming in people's lives, the inherent difficulty of understanding climate change, combined with the limitations of direct personal experience, also contributes to the lack of public understanding.[40]

Unfortunately, the majority of Americans, regardless of political affiliation, lack confidence in the government's ability to do more to address climate change.[41] Many people in the United States are often "demobilized by feelings of isolation, hopelessness, powerlessness, and lack of public trust in government to effectively address the issues," according to scholars Maxwell Boykoff and J. Timmons Roberts.[42] This is unfortunate, but perhaps realistic, as mitigating the negative effects of climate change through economy-wide policies may prove challenging largely because of administrative and political barriers.[43]

Decline and Rebound in Public Concern

In general, climate change and environmental issues in the United States have consistently ranked at the bottom of public concerns according to various polls. This problem is not limited to the United States, as opinion polls in several countries have observed a decline in climate concern and a corresponding increase in climate skepticism.[44] Concern about climate change seemed to peak around 2007 and then fell considerably afterward while simultaneously an increase in skepticism began to grow, even though it began before 2007. Interesting enough, the emergent global economic recession beginning in 2007 coincided with a downturn in climate change coverage, partially due to political and economic concerns.[45] The trend appears to support the notion that as quantity increases, so does public concern to environmental issues, while as the quantity declines, so do audience worries.[46] This is extremely troublesome with environmental reporting, largely due to the relationship between coverage and congressional action. For example, when rainforest, ozone, and GHG coverage increased during the late eighties and early nineties, so did congressional action against gas emissions.[47]

Opinion polls have detailed a decline in public concern about climate change.[48] For example, 41 percent of respondents in a 2007 Gallup poll stated that they worried a great deal about climate change, but by 2010 that percentage dropped to 28 percent.[49] Another poll in 2011 found that only 1 percent found environmental concerns as the most important problem facing the country,[50] while an additional poll that examined what issues Americans felt should be top priority policies found that global warming ranked third to last out of 18 issues.[51] Moreover, most Americans find global warming not that personally important to them. One study found that 73 percent of Americans indicate that the issue of global warming is "somewhat," "not too," or "not at all" personally important to them.[52] However, even with a decline in public concern and news media coverage, other studies suggest distress about global warming has been increasing since 2008.[53] About 73 percent of US registered voters in 2015 believed global warming was happening. The same percentage of Americans believed this in 2018.[54]

Public understanding of global warming also varies by political affiliation. Between 2001 and 2010, the percentage of Americans who believed the effects of global warming had already begun is enlightening. In 2001, 49 percent of conservatives believed this, but by 2010 this percentage had decreased to 30 percent. For Republicans the data show the same drop, but a little more pronounced to 29 percent. Liberals moved from 67 to 73 percent during the same time frame, while Democrats increased from 60 to about 70 percent.[55] This time period showed a clear decrease in the percentage of conservatives and Republicans who believed in the effects of global warming, even though the science of global warming had been overwhelmingly clear during the previous decade.

A few years later, most Democrats believed global warming was happening and that humans cause most of it. A little more than half were "extremely" or "very sure" global warming was happening in 2018.[56] Even though only 47 percent of conservative Republicans thought global warming was happening in 2015, that was a 20 percent increase from just two years earlier. In terms of global warming being caused by humans, 56 percent of voters in 2015 believed this, with half of "liberal" and "moderate" Republicans thinking this, but only 26 percent of conservative Republicans thought global warming was mostly human-caused.[57] By 2018, 62 percent of Americans thought that global warming was mostly human-caused. Unfortunately, only 16 percent of US voters in 2015, and 20 percent by 2018, knew that 90 percent or more of climate scientists were convinced that human-caused global warming was happening. Only 4 percent of Republicans knew this.[58] By 2018, about 25 percent of Republicans believed climate change was a serious problem, as opposed to 82 percent of Democrats and 51 percent of independents. A large number of Americans (69%) want the government to do more to reduce the activities that cause climate change, while only 22 percent oppose such action.

However, there is also a recent trend among Republicans who believe economic growth is more important than implementing environmental regulations to address the crisis. For example, in 2007, 58 percent of Republicans said stricter environmental regulations were worth the cost, while 34 percent said they would cost too many jobs and hurt the economy. Ten years later, in 2017, the opinions reversed. Now 36 percent of Republicans say stricter environmental laws are worth the cost, while 58 percent say they cost too many jobs and hurt the economy.[59]

Besides political affiliation, there is also evidence of an ethnic and racial gap regarding environmental and climate concerns, which is also comparable in size to the ideological gap between liberals and conservatives and greater than the partisan gap between Republicans and Democrats. Here, nonwhite minorities express higher levels of concern compared to whites, while women and lower income individuals voice more concern compared to men and higher income members. Regarding climate change, minority groups, which are often more vulnerable and aware of potential environmental impacts, may be motivated by concerns that are less based on political orientations.[60] This is an interesting finding, which also seems to support data that the vast majority of African Americans tend to vote Democratic, as do most Hispanic voters.

The political affiliation poll data suggest that the Trump administration is at odds with many fellow Republicans and conservatives on climate change. Many within the party disagree with their leadership on the issue of climate change. It should surprise no one that Republicans are capable of processing evidence when presented or exposed to correct climate change information. Even though many Republicans have challenged climate change science, they are not a monolithic group of global warming deniers. Republicans consist of the largest group that has increased its understanding and concern about global warming.[61] One study found that when exposed to accurate information about the scientific consensus on human-caused climate change, both Democrats and Republicans increased their support for public action.[62]

Unfortunately, regardless of political affiliation, the United States still exhibits an unusually high percentage of climate deniers compared to the rest of the world.[63] To repeat data from one recent poll, about one-third of Americans believed that climate change was not primarily caused by human activity, and that only one in 10 understood that scientists overwhelmingly agreed on the nature and existence of global warming.[64] If the press's role in a democracy is to provide citizens with significant and accurate information in order to make an informed and active citizenry, these polls provide a mixed review of the news media.

Additional Positive Trends

Nonetheless, with much of the US public's understanding of climate change increasing, the news is not all bad. Whereas blogs may not provide the solution to the limits of mainstream media climate change coverage, other media appear to be helping. While traditional corporate, mainstream media have a larger megaphone and better ability to cross-promote and share information, alternative venues are working to counteract some of their limitations. Independent and progressive media are often cited as better examples of how media can frame issues that challenge the corporate media's interpretation of climate change.[65] They are "well-positioned to generate the kind of communication needed to challenge established power as a precondition for addressing environmental crisis," according to professor Robert Hackett.[66] *Indymedia*, *The Nation*, and *Democracy Now* are viewed as covering social movements fighting global warming much better than the coverage coming from the mainstream, corporate-owned media.[67] Even though these progressive media outlets are also not as likely to fact-check or analyze scientific claims, they are more aligned on this topic with the European press for accepting science. For the United Kingdom, the nonprofit *Guardian* newspaper provides substantial climate coverage. It recently updated its climate change terminology policy, changing "climate change" to "climate emergency, crisis or breakdown." "Global heating" is now favored over "global warming." The progressive press appears to take the scientific claims and information generated from collective global bodies seriously. A main differentiating point between these media outlets and their corporate media colleagues is that the former are less driven by profit and less dependent on advertising dollars. Nonmarket, nonprofit, and independent media appear better at discussing the climate crisis as they aren't as hampered by advertisers, consumerism, profits, and corporate influence.

Another positive is that most media, even the mainstream corporate media, eventually accepted scientific statements indicating that global warming exists. In this regard, many journalists stopped covering climate change as a scientific controversy.[68] It is now "questioned" less often and less likely to be "controversial." In addition, even though sports, celebrity, political, and crime coverage outweighs that of climate change, public awareness is rising worldwide.[69]

However, once this became accepted as mainstream knowledge, the appeal to discuss it by journalists diminished.[70] Once there is less of a "debate" on issues, coverage can decrease. Fortunately, after over a decade of pushing fossil fuel misinformation, journalists are now inclined to approach climate change science from a "weight-of-evidence" approach, where stories are supposed to reflect scientific consensus.[71] Another study found that stories about climate change in US newspapers are also more likely to be framed with science compared with Canadian and international newspapers.[72]

Even with the many challenges facing the news media's abilities to discuss and cover the climate emergency, it appears public awareness is increasing. This is a positive trend and one that is partially attributed to the more outspoken and urgent calls for action, along with more dire forecasts from scientists and concerned citizens compared to years ago. Protests are also helpful with increasing awareness. More people are also experiencing extreme weather patterns caused by global warming. However, what the commercial news media do, or don't do, with that information is crucial. Will they translate the urgency of the science into calls for the immediate and radical change that is needed to address the emergency? Are they capable of conveying and supporting the actions that would likely eat into their massive profits—or are they themselves too compromised with the system? The next chapter examines these and other questions that explain why the commercial news media behave as they do.

CHAPTER TEN

The Battle against Global Warming Is Also a Fight for Media Reform

The first and second chapters of the journalism section detailed numerous internal and external challenges that journalists face when they write about climate change. The decline in the number of journalists, combined with the influence of owners, editors, and advertising, has paralleled a decrease in overall climate change reporting. Journalistic norms, in addition to the influence of fossil fuel industry think tanks, have also helped perpetuate misinformation regarding climate change. These challenges have influenced news media silence, especially regarding their failure to connect extreme weather or other environmental phenomena with global warming. These news media trends, along with the commercialization of the media, the myth of the liberal media, the politicization of climate science, the cocooning of the internet, as well as problems within political and economic systems, also influence public opinion and action regarding climate change. These developments suggest journalists are under immense constraints, which negatively affect climate change reporting.

In this chapter I explain *why* the quantity and quality of climate science reporting is constrained and how it can be improved. Literature that describes commercial media structures as a "cultural apparatus" helps contextualize and explain why climate journalism is severely constrained in the United States. At the very minimum, the challenges surrounding climate journalism suggest this is a structural problem, as many tenets of the existing news media structure clash with realistic solutions to the climate crisis. This suggests that addressing climate change also means supporting media reform. Opening up the US commercial media system to entail more nonprofit, noncommercial, and public news media will begin to disconnect journalism from many of the same influences that are fueling the climate crisis, and thus begin to improve the quality and quantity of climate journalism.

This chapter begins by noting the commercial news media exist within what critical scholars refer to as a "cultural apparatus," which creates a conflicting context for climate journalism. Its role with consumption is highlighted as it clashes with realistic solutions to the crisis. The following section continues with this approach by looking at *access* to mainstream commercial media and what it means for global warming. After that, the importance of having a strong public news component within media systems is discussed because it affects the quality of climate change reporting. Describing the differences in climate change coverage between the United States and European news media highlights this aspect. It becomes clear that many European models are superior, even though European climate journalism is far from perfect. The differences between US and European legislation and legislators and how they affect public opinion is also discussed. Finally, the chapter ends by providing a myriad of suggestions that can improve journalistic coverage of the climate emergency. Most importantly, democratizing US media will help climate change coverage if the media system moves away from a predominantly commercial structure to one that is predominantly noncommercial, where a significant portion of journalism is funded as a public good, not dependent on advertising revenue or government manipulation.

Cultural Apparatus

US news organizations exist within a larger political economy of neoliberal capitalism that celebrates profit accumulation, free markets, and individualization and views nature as subservient to the human race.[1] From a critical political and economic perspective, commercial media industries are considered part of a "cultural apparatus" whose overall mission is to mostly serve the interests of capital.[2] The primary concern of this apparatus is to incorporate more areas into markets while boosting profits.[3] Moreover, the cultural apparatus reinforces and stabilizes existing social and power relations, both politically and economically,[4] which means it can challenge policies and processes—like socialist approaches—that contest existing capitalist social relations.[5]

For decades, critical scholars have expanded the context of social contestation regarding the role of the media by suggesting that the media in general serve dominant rather than universal societal interests.[6] Fossil fuel company representatives, combined with their investors and allies, make up much of the global ruling class.[7] The existing holders of social power naturally see their way of running the world not only as the best way, but as the only possible way. Since the highly concentrated corporate structure of US media are part of the elite, they tend to reproduce their view of the world.[8] Edward Herman and Noam Chomsky, authors of *Manufacturing Consent: The Political Economy of the Mass Media*, reiterate that the underlying purpose of commercial mass media is to serve elite interests that mostly benefit from policies supportive of neoliberal economics.[9] The commercial corporate media, which generally follow and support government policy in their reporting,[10] particularly regarding foreign policy, successfully operate within and internalize the hegemony of US power relations. News professionalism, combined with modern news organizational systems, also often reinforce and legitimate the existing status quo.[11] The press reiterate dominant meanings of life to maintain systems of order.[12]

Because the logical extension of climate change research questions free market and neoliberal economic philosophies through the immediate need of government regulation of industries, a "cultural apparatus" combined with a Herman and Chomsky perspective would view the commercial media as reluctant to discuss solutions to climate change that challenge neoliberal economic policies, as there are few issues more challenging to the interests of the global elite neoliberal order. According to scholars Aaron McCright and Riley Dunlap, the pursuit of "environmental protection often involves government action that is seen as threatening economic libertarianism."[13]

In this regard, commercial media in neoliberal political/economies challenge journalists' abilities to provide thorough, in-depth information about global warming and proposed solutions. Indeed, some view the US media system as hostile to the level of quality needed given the importance of global warming.[14] External political and economic pressure, media ownership, and institutional values among other variables influence low-quality climate coverage.[15] Moreover, commercial and market forces function as a sort of censorship within media systems operating inside these confines.[16]

This is because media exist within and are directly tied into some of the main causes of global warming, infinite consumption through growth economics and environmentally damaging profit-oriented production processes. These two aspects are intertwined with most products and services that are advertised through the news media. Ad dependence influences media to favor middle to upper income audiences and encourages media to treat advertised products and their interests more charitably, while creating a positive buying mood for readers or viewers to react favorably.[17] This creates a troubling conflict of interest between climate change journalism and advertising interests. Put in other words, because "neither owners nor advertisers are disinterested parties in the operations of media companies, the upshot of these two realities is, to put it succinctly, *a bias in media content* favoring capitalist private business . . . and private personal consumption," according to economist Eric Schutz.[18] Communication professor Robert Hackett notes it is no doubt "difficult to introduce a kind of journalism that is systematically opposed to consumerism and that challenges corporate power and priorities and the extractivist state and hegemonic media."[19]

This means the news media will be less likely to identify consumption or advertising as *social problems* that directly influence the climate emergency. According to scholar Trish Morgan:

> When the neoliberal state has become little more than an instrument of capital, even state or public sector media are compromised by this nexus. Worse still, the privatized media sector are willing participants in the growth imperative and waste economy, and not likely to critique consumption or promote responsible attitudes to waste, as they are dependent on other corporations and consumerism for advertising revenue.[20]

This means the media largely perpetuate a "business as usual" approach to consumption, where people are to shop their way out of economic recessions, even in the midst of an ecological crisis.[21] To deviate from the status quo can result in excommunication from mainstream corporate media, while "non-consumerist, non-invidious, cooperative, egalitarian and ecologically harmonious ways of living in human community are downplayed, ridiculed or ignored as non-existent, irrational, quaint or comic—or at best perhaps colorful but unfortunately unachievable," according to Eric Schutz.[22] Morgan adds that "the mainstream media and cultural industries are unlikely to supply discourses of alternative thought that prioritizes well-being over consumerism, sustainability over conspicuous consumption, and downsizing over depletion and waste."[23]

Directly addressing any of the human activities and inventions that influence global warming, like the use of fossil fuels and capitalist growth-economics, challenge extremely powerful interests in society. As Jennifer Good, professor at Brock University Canada, tells us, "If the articulation of causes and consequences of climate issues can be threatening to those with a vested interest in maintaining the status quo, discussing solutions is potentially even more so."[24] With US commercial media existing within a capitalist society that is also the preeminent global superpower, its media and elites have perhaps the most to lose economically and politically from changing the status quo. Effectively addressing the climate emergency requires upending the current situation, and many in the US government resist this change. Commercial control of media, with its inhibited climate change coverage, helps perpetuate the delay and lack of government regulation needed to appropriately address the crisis. In essence, the media giants will most likely portray the world in such a way to perpetuate the political and economic status quo for those who have the greatest stakes in global power relations.

Regulating Public Access to Media

Voices that question consumerism or corporate power are often difficult to find in the mainstream corporate media. This raises the important question of who has access to the media. Access in this sense is about who is able to contribute to and publish in the mainstream corporate media. Access is also about exposure. A limited public ability to access the media challenges the promotion of proposed climate crisis solutions, which also challenges the quality of publicly accessible global warming information. Private media organizations can pretty much do whatever they want regarding who is allowed in their papers and in their radio and television programs. Since President Ronald Reagan deregulated the US media in the 1980s (or reregulated them in their favor), media owners do not have to provide any sort of semblance of equal time for differing sides. They have the individual right to freedom of corporate expression, which can mean legally marginalizing one perspective or artificially inflating another. They also have the right to allow owners' financial interests to influence newspaper editorials, as well as regular news reporting.[25] In this regard, the profit-maximizing orientation of huge media conglomerates can impede the mass media from fulfilling their role in providing space for public deliberation within democratic societies.[26] This means that the commercial media do not constitute anything similar to a "public sphere," or an area where concerns and problems are debated in an area free of government, commercial, or private/corporate influence. According to Christian Fuchs, "The public sphere is simultaneously process and space. Social organization turns into a public sphere when people act politically in common for a joint goal that fosters participatory democracy instead of economic and state power and when they use grassroots organizations and/or the occupation of public space as political strategy."[27]

In short, corporate control of the mainstream news media means control. They have the legal authority under property rights to run their media how they want to. The media access door is difficult to open for regular people, as the majority of media sources stem from elites. Several studies indicate that civic challenges to corporate-led globalization, which influences climate change, are most often negatively portrayed in the mainstream media.[28] According to Portuguese scholar Anabela Carvalho, "As political agents with positions, ideas, and proposals for addressing climate change and other public matters, citizens have been largely left out of media(ted) discourses."[29] Limiting civic groups' media access also negatively affects democratic potential.

US vs. European News Media, Public Opinion, and Legislation

News organizations in many countries around the world exist in structures that entail a much larger percentage of overall media that is considered publicly oriented and funded. Most European nations' media systems are structured with substantial percentages of media that are noncommercial and free from advertising interests. There continues to be a significant quality difference in media coverage of climate change between US and European news organizations. This variance also appears to influence the relationship between legislative rhetoric and news content, along with public opinion regarding climate change between the two continents. On all three counts, the United States lags behind Europe in terms of quality news media, public opinion, and legislative representation regarding climate change.[30]

US and European Climate Coverage

While chapter 8 compared the trends in content and quantitative output of US climate news reporting, noting how it was far behind compared to some European countries, along with Canada, this section examines mostly qualitative differences between US climate coverage and European news media. Media coverage in many European countries differs from US coverage in at least three ways. First, US media are regarded as having a much more pronounced upper-class bias compared to the European press.[31] This leads to coverage of issues and concerns that is generally directed toward wealthy business interests instead middle class, poor, or working-class interests.[32] Second, European coverage generally follows and supports the conclusions of the world's top climate scientists while, third, their reporting also tends to support political action in order to mitigate global warming. In both of these trends, US news media coverage falls well behind their European counterparts. In general, the US media display uncertainty regarding the science of global warming, while studies of various national media in Europe indicate the opposite, that there is scientific certainty regarding anthropogenic climate change.[33]

Another study confirmed this trend when it found that media coverage in Germany closely mirrored the scientific community as documented in the IPCC reports.[34] The German print media also call for political action in order to limit the damages of global warming, while their Swedish counterparts often write about global warming with a frame of certainty.[35] Swedish newspapers' focus on the European media's construction of certainty also helps legitimize the call for collective action regarding a European climate policy.[36] In this sense, Swedish media viewed themselves as having a responsiveness to their political setting, which manifested itself though a collective action frame of mitigation. The differences between the Swedish and US media are so extreme that a media outlet in Sweden noted in 2005 that the affluent part of the world is responsible for the changing climate and that these countries are now in debt to the developing countries. This type of approach to climate change is clearly missing from the US media,[37] as US public engagement on global warming lags well behind what is required for collective action.[38] The likelihood that the US commercial news media would make similar statements is very low.

However, one cannot say that Swedish, German, and other European news media coverage of climate change is perfect. Mediated climate change discourse in the United Kingdom[39] media can look confusing, contradictory, and chaotic, with many opposing voices.[40] In Germany, climate scientists who are very certain about anthropogenic global warming also perceive the news media as downplaying the topic, while presuming the media tend to nourish politicians' doubts about it.[41] Indeed, much of the press's analysis of climate change governance, especially in the early 2000s, remained within the ideological parameters of neoliberalism and free-market capitalism.[42]

According to Lee Artz, professor at Purdue University Northwest, "capitalist states encourage consumer atomization, privatization of social activities, and market rule: protecting the market and the reproduction of private property and capital, by enforcing wage labor and private profits."[43] This means that the press is most likely to avoid sustained critiques of growth economics and increased consumerism along with the international injustices associated elevated carbon emissions. In addition, American, Canadian, and international newspapers show a hesitancy to frame climate change with either oil reduction solutions or extreme weather consequences.[44] Additional climate change media frames also often inspire inaction.[45]

To further contrast the US news media with that of other countries, there is a troubling example that highlights the massive loss of partiality and some may even say integrity of at least three US media outlets. In 2016, the *Atlantic*, *Washington Post*, and *Politico* decided to host and moderate a climate-change deniers' event at the 2016 Republican National Convention.[46] These media organizations invited several deniers to speak, but environmentalists were nowhere to be found on the panels. The American Petroleum Institute, the lobbying arm of fossil fuel giants like Chevron, ConocoPhillips, and ExxonMobil, sponsor these US media-related denier events.[47] It seems quite odd for media organizations to treat climate denialism as a legitimate perspective when the evidence against it is so conclusive. While European news media colleagues may be scratching their heads in bewilderment with this case, it is not hard to see with examples like this why public trust in the US media is so low.

Superiority of European Climate News Coverage

Nonetheless, European news media systems that exhibit larger nonprofit public footprints do seem to have many positives associated with them. European scholars may not immediately realize this, but not only does their scholarship benefit knowledge about information systems in their own countries, it also benefits US-based scholars and citizens. By assessing models much differently than the US corporate-commercial model, European research highlights differences that are often more democratic than the US media system. These differences place many European media systems, especially the Scandinavian ones, at the top of international rankings regarding press freedom, quality of news coverage, and willingness to confront or question politicians.[48] The United States registers far below on all three counts. In fact, according to the 2016 World Press Freedom Index, the United States doesn't even register in the top 10, 20, or 30 countries.[49] It falls in at number 41, next to Burkina Faso, while Estonia, Namibia, Samoa, Lithuania, and Tonga, as well as a host of other countries from around the world, are listed as having a press that is more free than the United States. While the Scandinavian countries created press systems that are the most free, they are also among the most democratic countries the world offers, while their democratic-socialist governments contribute the most to public media compared with other nations. Unfortunately, the United States is no longer a fully democratic country and now constitutes a flawed democracy according to the conservative *Economist* magazine's democracy index, as noted in chapter 7.

Legislation Polarization

Regarding legislation, the United States also exhibits the strongest degree of climate polarization in legislative discourse compared to the United Kingdom, Germany, and Switzerland. About 63 percent of US legislative contributions are from advocates who want to address the crisis, while 37 percent are by skeptics. For Germany, the ratio is 75 percent advocates to 25 percent skeptics, with the United Kingdom somewhere in between. The German government recently created a goal to cut all GHG emissions by 80–95 percent by 2050,[50] while US president Trump does not believe in global warming and has refused to participate in global climate change accords. The United States also exhibits the most extreme skepticism in its legislative deliberations, with 18 percent, followed by the United Kingdom at 11 percent, and Germany at 3 percent.[51] It is important to consider that whenever Republicans vote against important environmental legislation, public perceptions about the threat of climate change adjust downward.[52]

International Public Opinion

The flawed nature of the US political and media systems in turn influences public opinion polls that indicate the US public is less aware about climate change than many other countries around the world. For example, the Nielsen Company found that Europeans and South Americans were more aware and concerned about climate change, while North Americans were the least aware and concerned.[53] In Europe, Germans retain a high level of climate change awareness at 70–79 percent, as do citizens in Sweden, Norway, France, and Finland.[54] In Europe, "outright denial is rare, but citizens' perceptions of climate change varies over time," according to one study.[55] In the United States, denial is more common and public understanding of climate change lags behind the science.[56] The United States has the third highest percentage of climate deniers in the world according to a 23-country survey, trailing only behind Saudi Arabia and Indonesia.[57]

Crisis of the US Media Model

With the evidence challenging the environment in which climate journalism can flourish piling up from the three chapters in the journalism section, one begins to question whether the mainstream US news media are structurally able to address major challenges facing humanity, such as the climate emergency, or if they have been too compromised. Obviously these chapters argue that the current structure of the US news system challenges the ability to convey the high-quality climate change news needed to adequately address global warming. Its limited nonprofit and public/community news infrastructure is not adequate enough to address these challenges. Efforts are needed to immediately expand this infrastructure if Americans want to improve US climate reporting.

If one must point the finger of blame, then the profit-oriented, commercial, corporate-dominated libertarian system of the US media must be reformed. It is a system that fits quite well with what the billionaire Koch brothers advocate: a privatized, corporate-led system with little government regulation and little-to-no citizen oversight, access, or control. According to professor Christian Fuchs, "There are power differentials in commercial media that disadvantage individuals and groups that do not have significant shares of money, political influence, and reputation. These structures disempower citizens' voices and visibility."[58] For-profit and advertising logic makes media organizations dependent on market and commodity logic, which makes them more prone to exclude voices that question these logics. Finally, there is an educational and economic gap that can privilege educated and wealthy individuals. In short, this system is not very democratic and it negatively affects the quality of climate change coverage.

Libertarianism

The mainstream US news media exist and profit in the same market-centered libertarian philosophies that the fossil fuel companies advocate and protect at nearly any cost, including the potential irreversible heating of the planet. The mainstream media's incentive to embrace regulation of the fossil fuel industry to address climate change counters its own concern against any public-interest government regulation of the media. Not only do fossil fuel interests help fund the media through advertising, but the media also share a similar philosophy opposed to government intervention/regulation of their industries because they can harm profit rates. Regulations that could significantly affect consumption patterns for certain groups within societies, in particular the wealthy, clash with the mainstream corporate media's philosophy of market-centered libertarianism. They also clash with the media's main source of revenue—advertisers. In general, it would be more strategic for the mainstream corporate news media to be united with the fossil fuel industry in opposing public-interest regulation of their industries. By supporting the regulation of another major industry—on which the media are partially dependent through advertising—the media open themselves up to the possibility that regulating a major industry in the public interest can be beneficial. This means that regulating the news media industry, perhaps by including a significant percentage of public media, could also be an option. This would also admit that a similar regulatory approach, or reenvisioning of the media system, would likely be necessary for better climate coverage, which ultimately means a smaller market share for the mainstream media as a significant nonprofit sector would be needed to begin to democratize the US media.

Time to Be Impartial or Take a Side?

Fear of offending and potentially losing certain members of the public, limited public access, and pressures from investors and owners for higher ratings and profits, among other constraints, means the mainstream news media will be hesitant to take a forceful stance in favor of immediate and far-reaching government regulated mitigation measures. In many cases, the US news media have been complicit in terms of either supporting atrocious US events, such as the invasion and occupation of Iraq, or have often been complicit by either omitting information or by spreading industry or government disinformation for a number of important topics (e.g., wars, oil pipelines, US coup d'états, nuclear power, and so forth). These are major issues that affect humanity, and for many are life and death situations.

Global warming, like propaganda about Iraq and other recent wars, goes through a commercial system that is not structurally able to adequately vet and address major crises. For the US invasion of Iraq, the US news media were a significant threat to peace as they were easily co-opted and conquered by government propaganda.[59] In this regard, the US news media were ineffective and partly responsible for either the damage incurred by these policies or by the lack of action. The damage from the US invasion of Iraq was extreme, while the damage to the Earth and humanity from global warming is predicted to be much worse in the long run. Moreover, it is telling that the media are interesting in discussing the potential economic downsides of climate change mitigation regarding business interests, but were unlikely to discuss the substantial public debt that would incur as a result of the Iraq War and President George W. Bush's Global War on Terrorism.

There is a similar situation where during the civil rights movement, some southern journalists began to believe that African Americans should be treated the same way as whites. There was pressure from wealthy, powerful conservative groups against this. The courageous reporters received a lot of flak for any sort of positive civil rights coverage, but many of them persevered covering the topic anyways. This is when the modern myth of the "liberal press" was born. It goes back to the 1950s when reporters began to display sympathy for the idea that African Americans deserved equal rights with white people.[60] The racists who opposed this began to label the media "liberal" as a result of their reporting. Even while this happened, many prominent journalists and newspapers supported racial segregation. Nonetheless, many in the media took an ethical, valued stance in support of equal rights.

While the Earth can't verbally speak to the population and explain its side of the story, it is however sending a message that it is rapidly deteriorating due to human activities. As qualitative social science tells us, science must be objective, but should not lack values. Within a news media context, it is good when news media are objective, but perhaps not so great when the news lacks values. In this case, the news media should immediately and strongly advocate for government-regulated mitigation in order to address the climate crisis.

Confidence in News Media

Activating large quantities of citizens to advocate for news media reform may be challenging because many have lost confidence in the US news media. However, the loss of confidence may help activate groups to reform it. Unfortunately, the lack of public access, the overall declining quality of news media during the past few decades, and corporate libertarianism have affected public confidence in the news media. This means it needs to change dramatically. Since 1985, there is an overall downward trend where Americans are becoming less and less likely to indicate that news organizations get facts straight and that media deal fairly with all sides. The vast majority of Americans (at 63%) believe news stories are often inaccurate, while just 29 percent of Americans say news organizations generally get the facts straight. By 2009, only 18 percent of Americans thought the press was dealing fairly with all sides of an issue.[61] In addition, journalists and the media continually receive bottom of the barrel ratings with public opinion polls indicating only about 28 percent of the public views journalists as contributing a lot to society. About the same percentage sees them contributing "not very much" or "nothing at all" to society.[62] Another poll found just 6 percent of the US public have a lot of confidence in the media.[63] Close to the 2016 presidential election, only 18 percent of Americans indicated they had high trust in the national news media. These numbers are quite different compared to 1972, where 72 percent of Americans expressed trust and confidence in the press.[64] Democratizing news media, along with addressing many of the challenges journalists face, will begin to improve the public's confidence in news media.

Improving Climate Change Journalism

There is an urgent need for truthful, as well as vigorous and engaging, journalism on climate science and policy-making.[65] Advocating for government-regulated mitigation similar to the European press would be a good start in terms of improving US journalism's coverage of climate change. While this may be the most immediate and relevant need, the challenges facing climate change journalism can be addressed through a myriad of suggestions. If editors agree, journalists should no doubt contemplate the results of their standard modes of operation via climate reporting.

They may want to reassess their practices given the outcomes and consider some alternatives. Better reporting practices, better training in science communication, broadening sources, avoiding false balance, and providing solutions may go a long way. Moreover, greater scientific literacy among the public and policy-makers, better funding for communication initiatives from governmental and research organizations, and more appropriate journalistic norms can also help public understanding of climate change.[66] One must also ask the question of whether journalists should omit climate deniers and skeptics from their coverage because of their false claims and misinformation. This is a tough question, but the answer is no. Everyone deserves a shot at accessing the media in order to communicate their ideas to the public. However, journalists should continue scrutinizing them, and give them less exposure, obviously much less than the "balanced" or 50–50 take on reporting, especially when the majority of scientists view their ideas as rubbish.

In addition, journalists may want to make a habit out of indicating where their sources' funding comes from, and if they are credible within scientific circles, while noting what organizations, industries, or companies the sources are affiliated with. Clarifying and labeling sources for scientific training, among other identifiers, can help.[67] Journalists also need to be careful with their reliance on official sources, which often limit the diversity of perspectives.[68] Moreover, journalists are also part of the elite,[69] and they experience pressure to conform with the behavior of other journalists, which influences the presence or absence of information and sources in mainstream media. This "pack" prevalence is often found in elite news agencies.[70] In the context of disasters, journalistic use of official news sources is the least likely to raise critical concerns about causes or consequences.[71] Indeed, journalists use their senses of what they feel is important for audiences as well, and elites are not the sole drivers of what comes out in the news. In many cases a circular process involving journalists, public opinion, and elites influences information. Clearly, journalism schools will need to emphasize the importance of including a wide variety of sources even if elites and other journalists are not referencing specific topics.[72] Including more foreign sources in reporting, in order to bring in voices and perspectives that deepen and extend political debate nationally, is also useful.[73]

However, it is extremely unlikely that these aspects alone can address the challenge of reporting on climate change. If journalists are better educated on the importance of diverse sources and their editors and owners allow this diversity, advocacy and activist groups will also need to reach out to the media more often by supplying information. This is another target area that can improve climate journalism. Perhaps environmental groups and scientists are not providing the media with enough of their studies. If journalists expand the diversity of their sources, then it should follow that the public would receive more representative sampling of perspectives. Indeed, public groups can offer counternarratives to corporate perspectives by advancing common sense approaches to energy choices by focusing on conservation, efficiency, and the further development of renewable sources.[74]

Moreover, changing frames that appeal to people by engaging the public in ways that give them agency to be politically active and influential may also be beneficial. These frames and others may need to be researched in terms of their potential effectiveness in support of creating an active public,[75] which is another area where university research can help. This is a tough challenge as media do not always motivate or have an awareness-heightening effect. The usage of informational media "only has a significant positive influence on everyday climate-related behavioural intentions when the actions are quickly realizable, have an individual economic benefit, or when it involves having an influence on politics," according to a recent study.[76] Therefore, it may be important to emphasize the implications of climate science in people's lives and that these implications require political action. The news media may also want to connect consumption, capitalism, energy use, political structures, and economic growth to the climate emergency, while also writing about overconsumption, sustainability, and divesting from the fossil fuel industry. Last, discussing more of the root causes of, and the solutions to, the climate crisis can improve coverage.[77]

However, changing these frames while expanding the topics may be more challenging than expected because they are often cultivated and defined by the political process, and journalism does not operate completely separate from the political system. Scholars find that media coverage of climate change reflects the mood coming from governments, which means that if governments take the issue seriously, the media are inclined to follow suit.[78] They are interconnected through sourcing and the idea that they must report on political activities. Often the political activities or nonactivity, if reported, become the generators of information for the public.

This means that the diversity of the political spectrum and the funding of the political system also indirectly influence coverage of climate change.

In terms of diversity, the US two-party system constitutes a political crisis of sorts. While many Republican representatives are major barriers to mitigating global warming, the Democrats have not made this a major priority either. Most other countries have more than two significant political parties, which often exhibit drastically differing views regarding actions required to address global warming, which can indirectly alter the makeup of news media content. When the political spectrum lacks diverse views on existing political issues, combined with a funding system supplied by the wealthy and major industries, these variables often limit the diversity of climate coverage and action. With the US political system dependent on money from these groups, it is unlikely that the majority of politicians from either party will dramatically alter the economic system, and reforming this system is required to help address the climate crisis. Eliminating money from US politics will help democratize the United States and help free political representatives from moneyed interests and influences. US media corporations consistently lobby against this (e.g., lobbying against reforming the *Citizens United* decision, which allowed unlimited donations to political campaigns) because it would likely lower their profit margins if political advertisements were limited or eliminated.

Given the structural constraints that challenge the US news media's ability to promote the action needed to address the climate crisis, applying pressure by protesting news outlets can also help improve the obligations that go with conveying the climate emergency to the public. For example, in 2019 in New York the climate activist group Extinction Rebellion briefly blocked a major avenue to the *New York Times* to protest what it viewed as the paper's inadequate coverage of the climate crisis. Hundreds of supporters called for an increase in coverage, more effective coverage, as well as released a "Climate Media Standards" guide for the media. The police arrested 70 of the activists. The Climate Media Standards guide[79] makes the following demands:

1. CLIMATE HEADLINES DAILY.
   - The climate crisis is the biggest story in history, and should be a front-page headline every day.
2. CLIMATE EVERYWHERE: EVERY BEAT, EVERY STORY.
   - **Discuss climate breakdown** in stories that already cover its causes and effects, and in politics, business, tech, world affairs, energy, immigration, health, real estate, food, sports, the arts, etc.
   - **Provide mandatory trainings for reporters and editors** on all

beats to understand the essentials of climate science, climate justice, the solutions available for the current crisis, and the history of global climate and ecological injustice.
   - **Report the links** to climate breakdown in every story about extreme weather and its effects. Cover the carbon count and include it in the daily weather report.
3. SUGGEST REAL SOLUTIONS.
   - **Stop** suggesting consumer-driven solutions.
   - **Start** suggesting community-driven solutions and mobilization. Acknowledge the necessity of systemic political and societal change.
4. USE CLIMATE EMERGENCY LANGUAGE.
   - **Stop calling it "climate change." Name the threat:** "climate crisis," "climate breakdown," "climate destruction," or "climate emergency."
   - **Use language that states the risks of the crisis more clearly.** For example, "global warming" becomes "global heating," and "biodiversity" becomes "wildlife."
   - **Do not publish statements that cast doubt** on the scientific consensus that a catastrophic climate crisis is under way and driven by human activity.
   - **Do not print** the opinion of writers who cast doubt on the scientific consensus and the effects of the climate emergency.
5. ENACT CLIMATE STANDARDS FOR POLITICAL ENDORSEMENTS.
   - At a minimum prerequisites should include support for halting all new fossil fuel infrastructure, support for a climate plan with zero emissions in 10 years or less and the No Fossil Fuel Money Pledge.
6. REMOVE FINANCIAL CONFLICTS OF INTEREST.
   1. **Ban advertising** from companies that trade in fossil fuels.
   2. **Divest** from fossil fuel companies.

These are excellent suggestions that will have a positive impact on climate reporting. Given the larger structure of the US media ecosystem that this book discusses, citizens and activist groups may also want to focus on changing the structure of the privatized, for-profit news media system that is largely funded by advertising interests that promote consumption. Climate change journalism is unlikely to change dramatically until the United

States democratizes its media and news information systems by creating a substantial sector free from market pressures, advertisers, and corporate ownership. A fundamental shift toward more publicly owned and operated news media free of commercial interests will begin to positively address some of the challenges described in Extinction Rebellion's Climate Media Standards guide.

Therefore, addressing the climate emergency also means media reform. Adding media reform into repertoires, changing media advertising systems, the continuation of protesting media organizations, creating additional strategies that apply pressure on news outlets, as well as by supporting media reform groups like Freepress.org, will help news media organizations more accurately address the climate crisis. Moreover, declaring a national emergency while creating a partnership with climate scientists, communication scientists, and news media organizations to routinely present mediated measures that are needed to address the crisis will also help public deliberation and action.

Democratizing US Media

This chapter points out that the crisis in the structure and funding of the US media means that the commercial nature of the US media system challenges journalists' abilities to report on climate change in ways that begin to drill down into the larger structural issues of consumption and sustainability. There is an obvious conflict of interest in any reporting regarding the role of consumption as the mainstream media are largely funded by advertisers dependent on consumption for their profits. The system by which media are able to exist is less likely to promote a different economic structure that could potentially diminish their market share. While contradictory and alternative views do exist within this system, the media's reliance on advertising in consumer-oriented economies suggests they would be more likely to propagate and support economic growth that surpasses the limits of nature. This means that the fight against climate change is also a battle for media reform. For climate change reporting to improve, there needs to be radical structural change within US media and news systems.

One way to address these US media system limitations would be to move away from a predominantly commercial system to one that is predominantly noncommercial, where a significant portion of journalism is funded as a public good not dependent on advertising revenue or government manipulation.[80] This would mean reforming, or democratizing, present US media policy. Democratizing the US media will most likely not just improve climate change coverage, it should also provide better attention to other issues as media systems with a significant percentage of nonmarket, public media generally exhibit higher quality journalism.[81]

Victor Pickard, one of the leading global scholars on the relationship between media and democracy, offers multiple solutions to broaden and fund noncommercial media. In his excellent book, *Democracy without Journalism? Confronting the Misinformation Society*, Pickard offers multiple funding paths to help reenergize news media.[82] His goal is to significantly increase public media while disconnecting them from advertising funding, which is an unreliable mode of support. In essence, the idea of a privatized business model coming to save journalism is delusional, as "no new business model that can save journalism is waiting to be discovered."[83] He notes that the results of the predominantly privatized US commercial news system yield excessive commercialization, misinformation on social media platforms, sensationalistic news coverage, clickbait, and degraded journalism instead of informative fact-based policy related news, which all lead to the structural collapse of professional journalism.

Given that media subsidies are firmly embedded in US history, he lists a myriad of fixes including creating public tax vouchers, AmeriCorps style journalism programs, charging broadcasters fees to use the public spectrum, creating a public trust, establishing a small tax on cell phones, taxing Google and Facebook, transforming existing post offices and libraries into public media outlets, breaking up media monopolies, establishing some worker control within newsrooms, while changing some journalistic practices, among other suggestions.

A much larger non-market-based media sector needs to be created, at least 30 percent of the total market share, which could consist of community, public, and nonprofit media.[84] In addition, studying the best aspects from Scandinavian media, in addition to some South American countries like Uruguay, may also be beneficial. While pushing this kind of media reform will most likely be ignored and resisted by the mainstream media, advocates in alternative and social media will have to take the lead on this. Democratization proposals will most likely first result in a near total mainstream media blackout, only to evolve into a massive backlash of propaganda from the corporate media sector including false accusations of media censorship. But as more and more communication scholarship demonstrates, a predominately privatized, profit-driven corporate media landscape censors diverse perspectives that are needed in democracies. While a new system will not ban or eliminate the private, corporate sector, it would open it up with new community and public voices that have often been left out of mainstream media coverage. Their voices will also be more free from the grasp of fossil fuel and other industry influences. While democratizing the media will not solve all of the problems, it will be a good start and should enable a better debate on what to do next.

Conclusion

This book examines three thematic areas specific to mass communication: the advertising system, digital era/infrastructure, and journalism. Each has shaped and continues to influence global warming along with government responses to it. It is important to remember that successful environmental movements can creatively dismantle or reform institutional infrastructures that extend the planetary global warming crisis and the unsustainable consumption of nature. This means that *the battle against global warming is also a fight for media reform.* Until we see structural change in the regulation of communication policies and industries, advertising and the digital world will mostly continue their current course. Much of the corporate news media will continue being a barrier that largely resists the drastic and immediate regulatory changes necessary to successfully address the worst aspects of this crisis. This is because so many of the needed changes challenge the media's source of income, their libertarian philosophy, and the general status quo that is preferred by elites.

In this conclusion I reiterate the book's major themes, discuss recent trends that offer some hope, and conclude with some relevant future actions people can take as individuals and collectives. As I argue throughout the book, we shouldn't overlook the power of the advertising system in the struggle for a sustainable world. Directly regulating US business and industry in favor of sustainability initiatives will be extremely difficult given their influence over the government. Reforming advertising can indirectly accomplish a substantial amount of what the IPCC suggests is needed in terms of lowering harmful GHG emissions. This may sound strange given that advertising evolved to propel forward a new consumer society that would stretch the limits of the Earth's productive and regenerative capacity by encouraging the overconsumption of goods and services with environmentally harmful and unsustainable production processes. Moreover, the relationship between advertising and global warming carries an underutilized history making it the field's most powerful contributor to the climate crisis. So how could it actually do something beneficial under these circumstances? While it is most often viewed with negative agency, its power suggests it can be a useful tool in the fight against global warming.

Advertising interacts with the main mechanisms (i.e., production, consumption, and growth economics) of human-made climate change in many ways. Understanding how they work together helps us better explain the causes of the climate crisis while laying a foundation for new policies to assist in reversing the crisis's trajectory. This book argues that the way advertising interacts with these mechanisms suggests they are *interdependent* to such an extent that altering anyone of the four mechanisms would change the trajectory of the climate crisis. This suggests advertising is a powerful influence in global warming and that, if modified, it would have a significant positive mitigative effect on the climate crisis.

To follow through with these arguments, this book rethinks advertising to present a mitigative model for environmental sustainability in hopes of sparking a larger conversation about how advertising can help mitigate the crisis. The macro effect of advertising is extremely powerful. If reformed, advertising can incentivize manufacturers to produce more environmentally sustainable products that cause less harm to the Earth's ecosystems. It is a gate that can allow, negate, or create fees for products and services depending on their relationship with sustainable production processes. This suggests that environmental and political groups may want to expand their mitigative strategies to include reforming the advertising system.

The second theme of this book focuses on the digital era and how it undermines efforts to address the climate emergency. The digital era has entailed a shocking growth in online data collection, e-commerce revenue, advertising, and overall consumption since the late 1990s. With the privatization of the internet, businesses could now collect and sell user data to advertisers, which enabled them to more effectively communicate to potential consumers with highly informed persuasive messages. This surveillance ability propels consumption forward while simultaneously exploiting online users and eroding their agency. The devices and software people use to access the internet are also purposely designed to become obsolete relatively quickly, which expends unnecessary natural resources. In addition, the amount of energy required to power the digital infrastructure is also unsustainable in the long term. Finally, the digital era, which exists within a neoliberal economic system, heightens preexisting inequalities by enriching a small but significant amount of tech workers who contribute elevated carbon emissions through their consumption practices. Their numbers also enhance concentrated power and political influence, which adversely affect democratic governance. All of these digital era trends negatively impact efforts to address the climate emergency.

The book's third theme focuses on the relationship between journalism and the climate crisis by discussing several US-based trends in commercial journalism that negatively impact global warming reporting. These developments challenge democratic processes and help delay political policy decisions urgently needed to address the climate crisis. Since the economic recession, many journalists have lost their jobs, which contributed to a drop in global warming news coverage and provided a wider opening for denialists to express fossil fuel misinformation. The journalistic norm of "balance" contributes to false balance, which often biases news content in favor of fossil fuel industry misinformation. US commercial news media continue to regularly engage in "climate media silence," where they ignore, downplay, or don't connect weather patterns with global warming processes. Writing about climate change in large corporate news organizations also clashes with advertising interests, media commercialization, profit orientations, editor preferences, and the owners' political orientations. Exogenous influences, such as the myth of the liberal media, the politicization of climate science, the cocooning of the internet, general inequalities, and disenfranchisement, also influence how citizens view climate science. Even with these challenges, recent data suggest the public is slowly becoming more concerned about the climate.

With the US media owned by wealthy corporations and run by elites who have the most to lose in a carbon-conscious world, news media organizations are part of a "cultural apparatus" that helps explain why climate science reporting is severely constrained in the United States. Moreover, the news media are also predominantly funded by companies that promote and benefit from overconsumption, which is one of the main causes of the climate emergency. The news media's profit orientation combined with a libertarian laissez-faire framework further challenge the ability of the population to reach a consensus in order to adequately address this challenge.

The sum of the endogenous and exogenous variables that challenge US journalism's coverage of global warming suggests this is a structural problem. Too many news organization modes of operation clash with what's needed to address the climate crisis, and given the urgency of the crisis, reforming the news media structure is urgently needed. Many European news media structures have a much larger publicly supported news infrastructure, free from advertising and commercial interests, which partially explains why their global warming coverage is superior to their US counterpart. While this doesn't fix everything, it opens up a space that is free from many of the negative influences that challenge climate crisis discussions and solutions. This suggests that democratizing the US media system is needed to better fight global warming. Citizen access to diverse information is required for a democratic society and US democracy is in a deep decline. Creating new nonprofit, local, national, and noncommercial news outlets will begin to disengage several of the influences that challenge US global warming reporting. It would also open space for the news media to call for the kinds of immediate government regulation required to address some of the worst aspects of the climate emergency.

Given the finite window of opportunity to address the climate emergency, media reform needs to happen very quickly, within the next few years. Local strategies and strategies inside and outside of institutions, like educating and pressuring political parties to adopt media reform as part of addressing the climate emergency, will help. Given the challenges at the national level, local strategies may be the key, especially with implementing advertising reform, which is the most powerful media reform tactic available in this struggle.

Needs and Action Summary

While the previous chapters highlight avenues where citizens can help address the mass communication problems discussed in this book, a quick summary of some of the main needs and citizen actions can help coalesce what's needed in the struggle for a sustainable future. In some cases, where the ideas or the field are relatively new (like mitigative advertising and reforming the digital world), raising awareness is the first step in creating change. In the case of journalism, a more historically established pathway for reform is already established. For advertising, there aren't any established groups people can connect with to help move this idea forward. This means people can assist by raising awareness about this possibility. This idea needs to get to other scholars and educators, as well as to environmental, activist, and political groups. They can further cultivate and debate the idea, contribute to its operationalization, lobby politicians, and spread the word. The more people contribute to a mitigative advertising model, the stronger it becomes.

There are more avenues people can participate in regarding the challenges to the digital world. The needs are clear: the creation of a nonprofit and noncommercial internet, breaking up monopolistic and oligopolistic tech markets, stopping price customization, and creating laws that allow individuals to retain their own online data. People also need to reevaluate how much technology is needed in their lives given the unsustainable energy requirements of the digital era. Manufacturers of digital items and software need to jettison obsolescence while opening up their supply chains to public scrutiny. Groups like the International Campaign for Responsible Technology, GoodElectronics Network, and Greenpeace are helping create policies that lead to a more sustainable digital future.

For journalism, it's well past time to create a substantial nonprofit and noncommercial news structure in the United States. Much of journalism needs to be funded as a public good, not dependent on advertising and free from government meddling in content. The previous chapter highlighted Victor Pickard's excellent proposals, which include public news subsidies, public tax vouchers, taxes on Google, Facebook, and cellphones, and a number of other creative solutions to the US journalism crisis.[1] Groups like Freepress.org are at the forefront of the movement to democratize US journalism, while *The Guardian* and *Democracy Now* are good places to learn about climate-related information. Media reform needs to happen immediately, not only to shore up a declining US democracy, but to better discuss and attend to the reforms needed to address the climate emergency. The planet's survival may depend on it.

Battling Back

There is progress being made in the battle for a sustainable future, which gives us hope. Progressive and democratic socialist politicians are putting forward legislation like the Green New Deal to address the emergency. The Sunrise Movement is also building bipartisan support for this congressional bill. Moreover, some media outlets appear to be improving, global protests are increasing, more groups are conducting climate actions, and much of public opinion now supports action.

In terms of news media, some media outlets are now beginning to listen to public pressure. For example, after months of insistence from progressive organizations and some Democratic candidates, CNN finally announced it would host a climate change town hall with 2020 presidential candidates. Moreover, MSNBC also announced it would cohost a multi-day climate change forum with the Georgetown Institute of Politics and Public Service and Our Daily Planet where 2020 presidential candidates from both major parties would be invited to participate.[2] In September 2019, the *New York Times* decided to end its sponsorship of the Oil and Money Conference, one of the world's biggest oil industry conferences, right after the Extinction Rebellion protests in front of their Manhattan offices. The *New York Times* has had a relationship with the event since 1980 and became an official sponsor in 2013.[3] Last, the *Columbia Journalism Review* and *The Nation* partnered in 2019 to create Covering Climate Now, which is a large-scale collaboration to increase climate coverage in the global media. The partnership included 250 newsrooms representing 32 countries with a combined monthly reach of about one billion people.

On January 29, 2020, the *Guardian* news organization initiated a policy to no longer accept advertising from oil and gas companies, which made it the first major news organization to effectively ban taking money from the fossil fuel industry. A few months earlier, the smaller but nonetheless influential Swedish *Dagens ETC* daily announced it would stop taking advertising that promoted fossil fuel-based goods and services. While this would be a financial hit for both media outlets, the *Guardian* decided it was the only logical thing to do when the world's leading scientists are telling humanity to change its behavior within about a decade if it wants to avert or limit a catastrophe. *Dagens ETC* decided this would be a wise decision in the long term.[4]

Besides these news media improvements, climate protests are becoming more forceful and effective, while their numbers around the world are increasing. Much of these protests are led by youth. For example, more than a million students from over 150 countries walked out of classes during the Global Climate Strike in September of 2019, with over 800 actions in the United States. Groups like Extinction Rebellion are also at the forefront of this battle. Moreover, additional youth protests sparked by Greta Thunberg continue taking place, often coordinated with her presentations in front of the UN and at other places around the world. Perhaps some of these actions helped lead the United Nations Secretary-General to finally call climate change an emergency in 2019.

There are now more groups around the planet supporting similar climate actions. In fact, there are over a thousand climate organizations around the world working to create a sustainable future. There are law, environmental justice, human rights, scientific, and a ton of other groups generating climate change advocacy projects and actions. For example, Climate Truth runs targeted campaigns to expose the fossil fuel denial industry, and has been successful in getting some corporations to drop their financial support for denial think thanks, like the Heartland Institute. Climate Truth also successfully mobilized public support in an effort to get major corporations like Google to withdraw from the conservative American Legislative Exchange Council, which influences legislation and denies the scientific consensus on climate change. Greenpeace USA also runs similar campaigns, often based on their PolluterWatch project, which investigates top donors like Koch Industries that contributed more than $100 million to 84 different climate change denying organizations during a 14-year period up to 2011. The Center for American Progress and the Union of Concerned Scientists also engage with the public about climate change denialism.[5]

These actions and protests, drastic weather changes, minor media improvements, and an evolving political climate may be affecting an increase in public awareness about the emergency in the United States. Many Republican voters now see the climate emergency as an important problem that needs to be addressed. Even in states like Louisiana that routinely vote Republican, 71 percent of voters believe in climate change and 50 percent say they're seeing its effects today.[6] According to another poll, 69 percent of Americans want their next president to take action, while 53 percent indicate action is needed right now.[7] In terms of action, 69 percent of registered voters, including 44 percent of Republican voters, indicated they support the Green New Deal, according to a recent study by Yale University.[8]

With increasing public support, resisting and creating plans for a sustainable future may become easier. Indeed, when faced with many crises, US citizens have often been able to confront crises and win in the past. Citizens helped end slavery, passed legislation allowing women to vote, and created laws regarding civil rights. The United States enacted the Environmental Protection Act, and it can't be forgotten that President Franklin Delano Roosevelt was able to reorganize the entire US economy to fight global fascism on an international level and win within a few years.

More recently, about 18 states raised their minimum wage, a Republican legislature in Tennessee enacted free community college, while Nevada expanded its voting rights. The American Recovery and Investment Act directed over $90 billion of public investment to low-carbon energy initiatives,[9] and teachers have gone on strike in many states to win crucial support. These recent achievements are occurring even as the United States continues to experience significant economic, racial, and gender inequalities. As one of the richest countries in the world, the United States has the resources, the people, and the abilities to confront many challenges—especially the global warming crisis. We have the capability to change things and anything is possible.

However, let's also be realistic. I do not want to be deceptively optimistic. Irreversible climate change is already occurring and will inevitably increase. The temperature of the planet will continue to increase; the question is by how much and that depends on what reforms we can accomplish. Make no doubt about it, there are tremendous challenges ahead in the struggle to address the climate crisis and this battle won't be easy. Politicians are trying to criminalize nonviolent resistance to the mechanisms and policies that propel the crisis forward. Moreover, the influence of concentrated power and money over political and communication systems almost has the system locked down in their favor.

While the corporate news media are finally acknowledging that global warming exists, their conflicts of interest will continue to negatively affect the coverage of potential solutions. For example, corporate news media suggest that climate change is not an existential threat, that the cost of action is too high, that addressing it is not logistically possible, that it's not politically feasible, that the American people won't support it, and that significant proposals (like the Green New Deal) that address the emergency will support government tyranny.[10] Efforts to reform advertising and the digital era/infrastructure will be met by a massive barrage of corporate and political propaganda, which the news media will then convey to the public because of their official source bias. Fortunately, media reform groups like Freepress.net are at the forefront of the battle to address these and other media challenges as they pressure the political system for media reform to help democratize US media. Nonetheless, the transition to a low carbon future will not come easily and it will require revolt, strikes, protest, and political participation in order to undercut and shut down fossil fuel capitalism and its supporters.

With this book I wanted to begin a discussion on new ways of rethinking the relationship between advertising, the digital era, journalism, and the climate emergency. Its optimistic praxis-oriented approach suggests that we will need to rethink how we interact with ourselves, political structures, advertising, the media, work, the economy, the environment, and our consumption habits. If we do not make these changes, much of life on Earth may not survive. Reforming advertising, the digital world, and the news media can be a part of a much larger link in a chain that can slow down and perhaps limit the worst aspects of the impending planetary environmental crisis.

Notes

Introduction

1. Epigraph from Sayers, *Creed or Chaos?* (1947), 47.

2. Mass communication is generally understood to consist of three parts: a message, a complex organization, and a heterogeneous audience. This process is where a complex organization sends a message to a heterogeneous audience. Dissecting the process creates a complex field where scholars examine a number of variables including economics, politics, ownership, ideology, effects, consumers, and messages. This book focuses on advertising, the digital era, and journalism, all of which fall under the mass communication field.

3. See Boykoff and Boykoff 2007. While some suggest that "global warming" is less accurate a term than "climate change," others suggest that "climate catastrophe," "atmosphere cancer," or "pollution death" may be more appropriate. See Corbett and Durfee 2004; Revkin 2008; and Storch and Krauss 2005. Culture contributes to perceptions of climate change: a comparison between the United States and Germany reveals insights about why journalists in each country report about this issue in different ways (Storch and Krauss 2005).

For this book, "global warming," "climate change," and various forms of "crisis" or "emergency," such as environmental crisis, ecological crisis, and planetary emergency are sometimes used. Professor Almiron notes that Frank Luntz, advisor to the Bush administration, pushed the term "climate change" in 2003 to replace "global warming" since it was less frightening. While much of the research I cite uses the term "climate change," a lot of it also uses "global warming." See Almiron 2020.

Given Almiron's research, I didn't want to only use "climate change" because I didn't want to make the climate emergency less threatening by omitting the term "global warming." At the same time I needed to give voice to the researchers I cite through their word use preference. Therefore, I use these terms interchangeably, similar to Boykoff and Boykoff (2007), who note they are used this way in policy and popular discourse. Some groups like Extinction Rebellion suggest using "global heating" instead of "global warming." Practically none of the documentation and research I cite uses "global heating" quite yet, so I used the term that's been used the most in climate-related research. I hope the overall purpose and argument of the book overpowers these details. This book is more about the crisis and emergency that surround global warming and the immediate need for mitigation.

4. See Stempel and Westley 1989.
5. See Chancel and Piketty 2015; IPCC 2014; OXFAM 2015.
6. See Chancel and Piketty, 2015; OXFAM 2015.
7. See Foster 2013; Hansen 2009.
8. See Hales et al. 2014.

Chapter 1

1. I am not a climate scientist, so I will let their studies speak for themselves in this section.

2. See American Association for the Advancement of Science 2014; Boussalis and Coan 2016; Hansen 2009; Karl and Trenberth 2003; Lewis and Boyce 2009. From a social science perspective, ecology as a scientific discipline developed to explain how economic and social-productive systems interact with the ecological systems in which they are embedded. Beginning in 1866, Ernst Haeckel coined the term "ecology" as an equivalent to Darwin's "economy of nature" concept, which focused on understanding complex plant communities (Foster 2015). During this period, Karl Marx drew on the work of German chemist Justus von Liebig and physician Roland Daniels to introduce the concept of "social metabolism," which connected human labor processes (society) to the environment (nature). Marx also attended lectures by British physicist John Tyndall, who discovered that carbon dioxide emissions contributed to the greenhouse effect. Marx eventually developed the theory of metabolic rift, which pointed to an irreparable rift in the interdependent process of social metabolism. Here, capitalist production processes treat nature as boundaries to be subsumed, eventually undermining the ecological foundations of human existence.

In succeeding decades, additional scientists further developed concepts of social metabolism, metabolic rift, and finite planetary space, which would evolve to solidify the foundations of a critical ecological perspective. In the 1880s, British zoologist E. Ray Lankester and botanist Arthur George Tansley followed Marx's work by creating a powerful ecological critique of capitalism; Tansley went on to develop the ecosystem concept in 1935. In 1926, V. I. Vernadsky explored the spatial limitations of production processes, noting that life exists on a limited and self-contained planetary sphere. By the 1940s onward, scholars became better armed with improved science, demonstrating how human action contributes to climate change and global warming. While there are a myriad of definitions and understandings of ecology, Foster (2015) may present the most salient and parsimonious definition. He writes that ecology "focuses on the human disruption of ecosystems from the local to the global" (3). This seems like an apt description of what is happening around the world with global warming.

3. See Wilkins 1993.

4. See Lewis 2013, 29.
5. See IPCC 2014.
6. See IPCC 2018.
7. See IPCC 2018.
8. See Watson et al. 2016.
9. See Goldenberg 2013.
10. See Steffen et al. 2018.
11. See Watson et al. 2016.
12. See Watson et al. 2016.
13. See Democracy Now 2018, n.p.
14. See IPCC 2018.
15. See Daly 1993.
16. See IPCC 2018 and Scheffers et al. 2016.
17. See United States Global Change Research Program 2018.
18. See Watts et al. 2018.
19. See InterAcademy Partnership 2018.
20. See United States Global Change Research Program 2018.
21. See Carrington 2018a, n.p.
22. See Leary and Lewis 2015, 31.
23. See Clark et al. 2016.
24. See Foster and Clark 2015, 3.
25. See United States Global Change Research Program 2018.
26. See Boykoff and Roberts 2007; Burton et al. 2002; Pielke 1998.
27. See Watson et al. 2016.
28. See Milman 2018a, n.p.
29. See Milman 2018b.
30. See United Nations Environment Programme 2018.
31. See Miller and Croft 2018.
32. See IPCC 2018.
33. See Watson et al. 2016.
34. See United Nations Environment Programme 2018.
35. See Anderson 2013.
36. See Sweeney 2014.
37. See IPCC 2014 and Oil Change International 2016.
38. See Jacobson and Delucchi 2009.
39. See U.S. Energy Information Administration 2014.
40. It is important to mention the risks of using nuclear energy as an electric power source. First, there is no satisfactory solution to nuclear waste disposal (Warf and Plotkin 1995). Second, the older waste is likely to last from 10,000 to 240,000 years, while newer waste poses a danger for 300 to 1,000 years and no geological burial system is likely to meet basic engineering requirements (United States Nuclear Regulatory Commission 2002; Warf and Plotkin 1995). Third, nuclear waste will also outlive the climate crisis if fixed (Stanton and Ackerman 2011), while manufacturing nuclear energy also results in up to 25 times more carbon emissions than wind energy (Jacobson and Delucchi 2009). Finally, regulation of nuclear power plants is often inadequate (Perin 2005; Union of Concerned Scientists 2012).

41. See Adams 2013.
42. See Anderson 2012.
43. See D'Alisa, Demaria, and Kallis 2014.
44. See Carrington 2019.
45. This is more so in the United States compared to Europe. For example, Europe has much better privacy regulations.
46. See Carrington 2018b.

Chapter 2

1. See Report: All major oil companies 2015.
2. See McCright 2006.
3. See OXFAM 2015.
4. See Carrington 2015.
5. See Lewis 2015.
6. See Associated Press 2018a.
7. See Zuboff 2019.
8. See Splichal 2009.
9. See Carvalho 2005.
10. See Carvalho 2005.
11. See Ewen 2001; Lewis, Inthorn, and Wahl-Jorgensen 2005.
12. See Park, Wang, and Pinto 2015.
13. See Jamieson 2014.
14. See Long 2016.
15. See Milman 2016b.
16. See McCright and Dunlap 2003.
17. See Boykoff and Rajan 2007, 209.
18. See Milman 2015a.
19. An activist group pasted posters with his picture on them during the 2015 United Nations climate talks. The group depicted him as one of many "climate criminals" who were wanted for "destroying our future" (Fountain 2016).
20. See Fountain 2016.
21. See Milman 2016a.
22. See Holden 2019a.
23. See Grandoni 2018.
24. See Holden 2019b.
25. See Venezuela in crisis 2019, n.p.
26. See Ward 2019.
27. See Soraghan 2018.
28. See McKnight 2019.
29. See Associated Press 2018b.
30. See Milman 2017b.
31. See Marsh 2017a.
32. See Milman 2015b.
33. See Goldenberg and Bengtsson 2015.
34. See Milman 2015a.

35. See Democracy Now! 2016.
36. See Milman 2017a.
37. See Marsh 2017b.
38. See Marsh 2017a.
39. See Friedman 2019.
40. See Trump admin escalates 2019.
41. See Democracy Now! 2018.
42. See Interior Dept. 2020.
43. See Lartey 2017.
44. See Milman 2017c.
45. See O'Neil 2019.
46. See Verburg 2016.
47. See Korten 2015.
48. See Harish 2012.
49. To be fair, Obama may not have been as bad compared to Trump. However, the Dakota Access pipeline protesters in 2016 experienced heavy law enforcement, FBI and Department of Homeland Security surveillance, and military style tactics used by pipeline security contractors during his administration (Federman 2020).
50. See Federman 2020.
51. See Brown 2019.
52. See Dodd and Grierson 2020.
53. See Nuccitelli 2018, n.p.
54. See Reports: EPA 2019.
55. See Democracy Now! 2019.
56. See Guardian Staff 2019.
57. See Cutler and Dominici 2018.
58. See Holden 2020a.
59. See American Lung Association 2019; LaMotte 2020.
60. See WHO says 2018.
61. See EPA indefinitely suspends 2020.
62. See Holden 2020c.
63. See Sneath 2020.
64. See EPA to introduce 2019. Besides creating environmental policies that would escalate the US death toll from the pandemic, the Trump administration had even more to offer the American people. The Make America Great Again Committee, a joint fundraising committee authorized by and composed of Donald J. Trump for President, Inc. and the Republican National Committee, sent out this in an email dated May 9, 2020:

"We have some exciting news to tell you. President Trump knows the past few weeks have been extremely difficult for Americans from all across the Nation. He appreciates your unwavering support during this time and wants you to know that it has not gone unnoticed. He wanted to do something special for you to show you how much your loyalty means to him, so he's asked us to give you **EXCLUSIVE ACCESS** to get our **Official Trump-Pence Pint Glasses**."

65. See Milman 2016b.
66. See BBC 2017; Horgan 2018.

67. See Wilson 2019, n.p.
68. See McCright and Dunlap 2011.
69. See McCright 2006.
70. See McCright and Dunlap 2011.
71. See Holden 2020b.
72. See Brulle 2000; McCright and Dunlap 2000.
73. See Boykoff and Rajan 2007. Even though the term "skeptic" is a good term and people should always think critically about information, in an environmental context the term takes on a new meaning. Jacques, Dunlap, and Freeman (2008) view the term as now describing someone who denies the seriousness of an environmental problem, dismisses scientific evidence proving the problem, and questions the relevancy of regulation to address the problem, while viewing environmental regulation and progress as competing goals.
74. See Matthews 2015.
75. See Hackett 2016.
76. See Kahan et al. 2007.
77. See Labor Network 2015.
78. See Prasad et al. 2016, n.p.
79. See Prasad et al. 2016. To be fair, the Western world developed its strong economies by contributing to the majority of the world's carbon and GHG emissions. Is it not fair to allow the so-called "developing" world the right to continue using fossil fuels to develop and build stronger economies on the level of the West? This suggests that the West would need to decrease its carbon and GHG emissions ever further in order to counterbalance any increase in emissions from "developing" countries. On the other side of the coin is to reject the notion that Western development should be the ideal for the rest of the world. The West should be moving away from growth economics as they have pretty much created the current crisis. Perhaps it's time to tone down the greed, overconsumption, and materialism that inflict most of the West. Instead, incorporating indigenous or other methods of living harmoniously from the "developing world" may help address the crisis.
80. See Symons 2019.
81. See Aronoff 2019.
82. See Aronoff 2018, n.p.
83. See DNC lifts ban 2018.
84. See Rosenberg 2019.
85. See Lacy 2019.
86. See Milman 2018.
87. The exception is the progressive wing of the Democratic Party, which has introduced helpful Green New Deal legislation. However, the mainstream media tend to marginalize this group.
88. Fortunately, there is traction within the US political system to address the climate emergency. Bernie Sanders and Alexandria Ocasio-Cortez have introduced a bill to declare global warming an actual government emergency. This would help jump-start the government's ability to address the crisis.

Chapter 3

1. John Galbraith, *The Affluent Society*, rev. ed. (1958; rpt., Boston: Houghton Mifflin, 1969), 149.
2. See Driver and Foxall 1984, 1.
3. See Goodwin et al. n.d.
4. See Pollay 1989, 446.
5. See Simon 1970, 194.
6. See Richards and Curran 2002, 74.
7. See Dahlen and Rosengren 2016, 340.
8. See Horkheimer and Adorno 2002.
9. See Dahlen and Rosengren 2016; Ellul 1965; Fuchs 2015; Jowett and O'Donnell 2006; Heilbroner and Thurow 1998; Media Education Foundation n.d.; Packard 1960; Richards and Curran 2002; Storey 2017.
10. Many scholars see green advertising as having a very limited effect in terms of mitigating the climate crisis (Fuentes 2014; Hamilton 2009; Kilbourne 1995). See chapter 5 for more on green advertising.
11. See Pearlin and Rosenberg 1956; Storey 2017.
12. See Press Association 2016.
13. See Carey 1989, 14.
14. See Carey 1989, 14.
15. See Carey 1989.
16. See Norris 1989.
17. See Comor 2008, 160.
18. See Andreucci and McDonough 2014.
19. See Dyer 1982.
20. See Rotzoll, Haefner, and Sandage 1976.
21. See Holm 2017.
22. See Marx 1973.
23. See Storey 2017.
24. See Morgan 2015.
25. See Baran and Sweezy 2013b, 38.
26. See Baran and Sweezy 2013b; Jhally 1997; Packard 1960.
27. See Northrup 2003, 319.
28. See Schudson 1989.
29. See Schudson 1989, 45. While some of this had to do with the massive promotional pitch to consume more and more, it is also likely that the consolidation of many products under one large national chain store was viewed as a threat to smaller, local businesses, which holds true to many today.
30. See Schudson 1989.
31. See Northrop 2003, 320.
32. See Bagdikian 1989.
33. See Schudson 1989.
34. See Corrado, Hulten, and Sichel 2009.
35. See Comor 2008.
36. See Schor 1999.

37. See Kanner and Gomes 1995.
38. See Zukin 2004.
39. See Goodwin et al. n.d.
40. See Saxton 2017.
41. See McKinsey and Company 2014.
42. See Goodwin et al. n.d.
43. See Dominick 2009; Jowett and O'Donnell 2006.
44. See eMarketer 2014a.
45. See Mad men are watching you, 2011.
46. See World Federation of Advertisers 2015.
47. See IPG Mediabrands 2014.
48. See Handley 2016.
49. See Comor 2008; eMarketer 2014b. This book has a purposeful US/Western-centric focus and often cites data from the United States. This is because the United States and the West are the most developed consumer-oriented capitalist systems in the world that consume the most manufactured commodities, use the largest percentages of the Earth's resources, and have historically contributed the most GHG emissions. Therefore, mitigation strategies should initially target these countries.
50. See Comor 2008.
51. See Comor 2008.
52. See Berry 2013.
53. See IPCC 2014.
54. See IPCC 2014.
55. See IPCC 2014.
56. Of course there are noncommercial public service announcements and advertisements encouraging people to read, among other forms of advertisements that do not promote a product or service, but they constitute the margins of commercial systems. The vast majority of advertisements promote purchasable experiences or material items.
57. See Fuchs 2015; Eagleton 1991.
58. See Storey 2017, 111.
59. See Scatamburlo-D'Annibale 2010; Storey 2017.
60. See Pearlin and Rosenberg 1956; Storey 2017.
61. See Ewen 2001; Lewis, Inthorn, and Wahl-Jorgensen 2005.
62. See Baran and Sweezy 2013a, 45.
63. See Foster and McChesney 2013.
64. See Morgan 2015.
65. See Brulle and Young 2007; Simon 1970.
66. See Cook 2011.
67. See Gunderson 2015.
68. See Held 1980.
69. See Pepper 2002; Gorz 1994; and Williams 2010.
70. See Kornai 1992.
71. See Beckwith 1949.
72. See Beckwith 1949; Gerth 2013; Hanson 1974.

73. See Beckwith 1949.
74. See Klein 2018, n.p.
75. See Ghotge 2018.
76. See O'Connor 1989.
77. See Asara et al. 2015; O'Connor 1989.
78. See Pineault 2018.
79. See Kallis et al. 2018.
80. See Kallis 2017, 2.
81. See Vergara-Camus 2019.
82. See Kallis 2017.

Chapter 4

1. See IPCC 2014.

2. Contributing to theory development that connects advertising with the climate crisis can aid in explaining Kaplan's (1964) "disturbing situation" theory requirement, and fulfill Greer's (1969) need for theory to be policy relevant, while further exploring the ways in which mass communication can be beneficial for society (DeFleur and Ball-Rokeach 1975; Tichenor and McLeod 1989). On a side note, transportation also expends significant GHG emissions and could be a separate variable in this model. However, I fold transportation into the overall production and consumption process since production and consumption are general measurements of economic growth.

3. The content of advertisements is irrelevant to this book. The transactional and distributive aspects of advertising constitute the main focus.

4. See Kotler et al. 2006.
5. See Dongling 1999, 1.
6. See Galbraith 1969; Hamilton 2009.
7. See Jhally 1997; Packard 1960.
8. See Galbraith 1969.
9. See Galbraith 1969, 152.
10. See Marx 1973.
11. See Brems 1951; Carey 1960; Galbraith 1969; Schumpeter 1934.
12. See Dyer 1982; Faraone 2011; McChesney 2004; Sukhdev 2012.
13. See Frank 2009.
14. See Bauman 2007; Ewen 2001.
15. See Faraone 2011; Sukhdev 2012.
16. See Becker and Murphy 1993.
17. See Tuchman, Nair, and Gardete 2015, 18.
18. See Simon 1970.
19. See Brulle and Young 2007, 539.
20. See Assadourian 2010, 12.
21. See Kallis et al. 2018.
22. See Asara et al. 2015.
23. See Vergara-Camus 2019.
24. See Kuznets 1973, 247.

25. See Asara et al. 2015.
26. See Comor 2008; Goodman and Cohen 2004.
27. See Horowitz 1994.
28. See Horowitz 1994.
29. See Morgan 2015, 71.
30. See Faraone 2011; Simon 1970, 171.
31. See Schumpeter 1934.
32. See Carey 1960; Galbraith 1969; Schumpeter 1934.
33. See Baran and Sweezy 2013.
34. See Faraone 2011.
35. See Simon 1970, 187.
36. See Simon 1970, 167.
37. See Yang 1966, 1–2.
38. See Simon 1970, 188–89.
39. See Brems 1951, 85–86.
40. See Ackerberg 2001; Ackley 1961.
41. See Brown 2001; Brower and Leon 1999; Gardner and Stern 2002; Jurado and Maldonado 2014; Storey 2017.
42. See Carey 1960.
43. See Dyer 1982; Hamilton 2009; Shapiro 2014.
44. See Galbraith 1969.
45. See Simon 1970.
46. See Galbraith 1969, 150.
47. See Galbraith 1969, 151.
48. See Kindleberger 1958.
49. See Kallis et al. 2018.
50. See Kallis 2017.
51. See Pineault 2017.
52. See O'Connor 1998, 1999.
53. See Kallis 2017.
54. Cited in Kallis et al. 2018.
55. See O'Connor 1998; Polanyi 1967.
56. See Harvey 2014; Agger 1987.
57. See Asara et al. 2015.
58. See Foster,2013; Foster, Clark, and York 2011; Klein 2014; Rockström et al. 2009; Shapiro 2014.
59. See Daly 1993. This is referred to as Daly's Impossibility Theorem.
60. See Daly 1993; Foster 2017; Klein 2014.
61. See Magdoff 2011, 2014.
62. See Alperovitz 2014.
63. See Foster 2013.
64. See Packard 1960; Sterne 2007.
65. See Berry 2013, 129.
66. See Klein 2014.
67. See Brown 2001; Brower and Leon 1999, Gardner and Stern 2002.
68. See Shapiro 2014, 243.

69. See OXFAM 2015.
70. See Chancel and Piketty 2015.
71. See Jackson2010.
72. See Berry 1995; Suzuki 1997.
73. See Comor 2008.
74. See Heede 2013.
75. See OXFAM 2015.
76. See Ge, Friedrich, and Damassa 2014.
77. See Ge, Friedrich, and Damassa 2014.
78. See Watson et al. 2016.
79. See Odum 2007.
80. See Sato 2012.
81. See Comor 2008.
82. See Antilla 2005; Boykoff and Boykoff 2007.
83. See Gardner, Assadourian, and Sarin 2004; OXFAM 2015.
84. See Hirschman 2011; Congressional Budget Office 2007.
85. See Goodwin et al. n.d.
86. See Reuters 2019.
87. See Shanahan 2007.
88. See Olausson 2009.
89. See McDonald 2015.
90. See Wackernagel and Rees 1996.
91. See Hackett 2016.
92. See Chancel and Piketty 2015.
93. See Chancel and Piketty 2015; Congressional Budget Office 2007; Comor 2008.
94. See Shah 2014; Worldcentric n.d.
95. See Democracynow 2018, n.p.
96. See OXFAM 2015.
97. See Maniates 2002.
98. See Chancel and Piketty 2015, 2.
99. See Schutz 2011, 178.
100. See Schutz 2011, 178.
101. See Jeb Loy Nichols, "To Be Rich (Should Be a Crime)," in *Long Time Traveller*, On-U Sound Records, ONULP130.
102. See US Energy Information Administration 2016.
103. See US Energy Information Administration 2016.
104. See InfluenceMap 2019.
105. The possibility that reforming or eliminating advertising could negatively affect demand and consumption within capitalism, perhaps tilting it into crisis, needs to be addressed. For example, allowing "environmentally friendly"/sustainable manufacturers to advertise would mean most environmentally harmful manufacturers would have to either change, or go out of business. This would be good for the environment, but may not be good for those who work in those industries. Baran and Sweezy (2013) propose that abolishing or limiting advertising would help address capitalism's negative consequences for consumers, values, and the me-

dia, but only if accompanied with full and socially desirable employment. They find that the elimination of present forms of advertising would most likely require the elimination of capitalism. Morgan (2015) points out that "if advertising channels were removed as sites of absorption, a significant portion of the waste economy would disappear, creating a problem for monopoly-finance capital and its stagnation tendencies" (76). A large decline in consumer spending would likely send sales and employment down, eventually resulting in a recession (Schor 1998). In this sense, if a capitalist economy is not growing it is in crisis.

Capitalism can go into crisis if the relationship between economic growth, consumer market capacity, and demand size breaks down (Baran and Sweezy 1966; Ryazanov 2015). Basically, altering or eliminating advertising can mean limiting growth and crisis, and a no-growth economy is not capitalist (Harvey 2014). However, Schor (1998) suggests that such an "apocalyptic scenario is unrealistic" (169), and that "retaining a healthy economy is possible if consumer spending slows considerably or even drops" (170). She argues that companies can shorten the number of work hours to keep people employed, which would likely increase their productivity. However, she omits that less work often means less money, and with many people struggling to make ends meet, this may not be a realistic solution. One could add that countries with full employment and a job security safety net, perhaps with a universal livable income, would be more effective in terms of supporting an advertising system that would account for sustainability.

The question is whether capitalism can alter itself before destroying the planet. Moving away from a capitalist economy would be more likely to support a more harmonious and healthy relationship with the planet. Even the IPCC finds decoupling "economic growth from human well-being" will assist in mitigating the environmental crisis (IPCC 2014, 288).

106. See Craig 2018; Fuchs and Linchuan-Qiu 2018; Just and Puppis 2018.

107. See IPCC 2014.

Chapter 5

1. See Craig 2018; Fuchs and Linchuan Qiu 2018.
2. See Brower and Leon 1999; Brown 2001; Carey 1960; Galbraith 1969; Jurado and Maldonado 2014; Morgan 2015.
3. See Murdock 2018.
4. See Baran and Sweezy 2013; Jhally 1997; Schutz 2011; Shapiro 2014.
5. See Daly 1993; Foster 2013; Packard 1960.
6. See IPCC 2014b, 288.
7. See Chancel and Piketty 2015; IPCC 2014b; OXFAM 2015.
8. See IPCC 2014b.
9. See IPCC 2014b, 83.
10. See Demaria et al. 2013; Kallis 2011; Latouche 2010.
11. See Kallis 2018.
12. See Jhally 1997; Shapiro 2014; Baran and Sweezy 2013.
13. See Rotzoll, Haefner, and Sandage 1976.
14. See Driver and Foxall 1984.

15. See Driver and Foxall 1984.
16. See Waterson 2018.
17. See Mulholland 2014.
18. See Hansen 2010; Hartmann et al. 2017; Shin, Ki, and Griffin 2017.
19. See Kilbourne 1995, 7.
20. See Kilbourne 1995.
21. See Lieberstein and Benjamin 2010.
22. See International Chamber of Commerce 2011, 2. It is important to note that this model in this chapter does not affect speech; all corporations will still have free commercial-speech under this model. The difference between companies is that those that do not meet the environmentally sustainable production practices outlined in the model will have a fee added on their advertising "speech." Companies with sustainable production processes would not. Although one could reframe that as discrimination, few would argue that fining a company for pollution is discrimination. That same concept applies to this model. Moreover, companies already pay different fees for their transactions based on their target, the medium of communication, the timing, and the type of advertisement. These variables could also be seen as discriminating speech in favor those who can pay. Those with the most money then have the best opportunity to present their speech.
23. See Hamilton 2009, 574.
24. See Fuentes 2014; Hamilton 2009; Kilbourne 1995.
25. See Budinsky and Bryant 2013.
26. See Kahn 2009; Maxwell and Miller 2012b.
27. See Maxwell and Miller 2012a, 31.
28. See Kahn 2009.
29. See Budinsky and Bryant 2013.
30. See IPCC 2014b.
31. See Fuchs 2015.
32. See Taylor 2014.
33. See Springmann et al. 2018.
34. See Nuccitelli 2017.
35. See Deeg et al. 2019.
36. See Harris 2018.
37. See Halperin 2018.
38. See Baran and Sweezy 2013.
39. See Lewis and Boyce 2009, 8–9.
40. See IPCC 2007.
41. See Lewis and Boyce 2009, 3.
42. See IPCC 2014b.
43. It is important to note that not all new renewable energy sources are completely harmonious with nature. Producing solar panels expends resources and wind power generators can kill birds.
44. See Apte, Karmarkar, and Nath 2007.
45. See Storey 2017.
46. See Chancel and Piketty 2015.
47. For more information on models, the following sources helped inform and

influence the book's proposed model: Magdoff's (2014) ecologically sound and socially just economy, Daly's (1993) impossibility theorem, Murdock's (2018) moral economy of machines for media materialities, and Jhally's (1997) discussion of advertising.

48. Turning down or banning advertising for any product or service with a significant carbon footprint would essentially drive the news media out of business given its reliance on advertising to fund its operations. In its discussion about rejecting advertising from cars, or fancy vacations, the *Guardian* noted "Stopping those ads would be a severe financial blow, and might force us to make significant cuts to Guardian and Observer journalism around the world" (Waterson 2020, n.p.).

49. See Daly 1993.

50. See Kilbourne 1995.

51. See Oil Change International 2016; IPCC 2014a.

52. See Packard 1960.

53. See Taylor 2014.

54. See Leonard 2011.

55. See Maxwell and Miller 2012a.

56. See Herman and McChesney 1997.

57. See Hardy 2014, 150.

58. See Laclau 2014, xii.

59. See Assadourian 2010.

60. See Gartner 2007.

61. See CDP 2016.

62. See Dangelico and Pontrandolfo 2015.

63. See Sheehan and Morrison 2018. However, that same study also found that professional organizations within advertising had little to share about climate, green, or sustainability issues. Moreover, industry ideas of addressing the climate crisis focused on doing creative work around ideas, while externalizing change onto citizens. Advertising professionals believed that the solution lies in convincing consumers to purchase environmentally friendly products, and that they could help convince consumers to consider how their own personal lives impact the crisis.

64. Cited in Jurado and Maldonado 2014, 117.

65. See Vaughan 2016; Jurado and Maldonado 2014; Wolonick 2012; Mosco 2014.

66. See Wolonick 2012.

67. See Mosco 2014.

68. See Jurado and Maldonado 2014.

69. See Mosco 2014.

70. See Vaughan 2016.

71. See Vaughan 2016.

72. See Just and Puppis 2018.

73. According to a study of US voters in 2015, Americans are very likely to vote for a presidential candidate who strongly supports taking action to reduce global warming, and less likely to vote for one who strongly opposes taking action (Leiserowitz et al. 2016). While only 39 percent of American registered voters think the

American people can convince Congress to pass significant legislation to reduce global warming, Americans still support a broad array of policies designed to reduce carbon pollution and dependency on fossil fuels, among other clean energy proposals (Leiserowitz et al. 2016).

74. See United Nations Environment Programme 2018, n.p.
75. See Happer and Philo 2016.
76. See Handley 2016.
77. See Kollewe 2017.
78. This assumes a nonregulatory government trajectory.
79. See Hern 2017.
80. See McChesney 2004.
81. See Hardy 2014.
82. See Artz 2017.
83. See Dominick 2009.
84. See Carrington 2018.
85. See Nuccitelli 2018.
86. See McChesney 2013, 208.
87. See Taylor 2014.
88. See IPCC 2014b.

Chapter 6

1. See Turow 2011, 5.
2. While this chapter focuses primarily on the United States, the consumer ramifications for data mining and individualized targeting have global implications. Different online pricing schema based on consumer surveillance are already established in Europe, with 40 percent of online retailers and mail-order companies in Germany offering variable pricing (LBBW 2017), while industry experts in the UK are predicting that fixed prices for consumer goods and services will be virtually nonexistent within a few years (Morley 2017).
3. Consumer sovereignty in the digital era implies an ability to know when, how, and why online data and price customization are being used for or against consumers, while agency entails the level of empowerment, or ability to influence, counter, or control how data is used and prices are formulated.
4. See Newman 2014; Stiglitz 2002.
5. See Negroponte 1996.
6. See Benkler 2006.
7. See Shapiro 1999.
8. See Rezabakhsh et al. 2006; Fisher 1998.
9. See some of this discussion in Mosco 2017. Also see Clark 1998; Danet 1998; Goldstein 1994; Pavlik 1998.
10. See Garcia 2015; Zuboff 2019.
11. See U.S. Government 2016.
12. See Srinivasan, Finn, and Ames 2017.
13. See Turow 2011.
14. See Pariser 2012.

15. See Nichols and McChesney 2013, 231.
16. See Kolko and Reid 1998; Sunstein 2001; Turow 1997.
17. See Splichal 2009.
18. See Schiller 2014, 1.
19. See Schiller 1996.
20. See Babe 1994.
21. See Melody 2006.
22. See Dyer-Witheford 1999; Ritzer and Jurgenson 2010; Schiller 2014.
23. See O'Neil 2017; Eubanks 2018; Vaidhyanathan 2018.
24. See Srnicek 2017.
25. See Bridle 2017; Taplin 2017.
26. See Browne 2015; Eubanks 2018; Noble 2018.
27. See Andrejevic 2011, 2015; Fuchs 2010, 2012b.
28. See Winseck 2016.
29. See Turow 2011.
30. As noted previously in this book, while not all advertising is effective, there is a historical macro relationship between the amount spent on advertising and overall consumption, with the increase in consumption paralleling amounts spent on advertising (Park 2017). For the digital world, a recent Comscore study described by Fuchs (2012a) found that people exposed to Facebook ads were more likely to purchase products online or in stores compared to those who are not exposed. The most viewed ads are those from the retail sector, which constitute 23 percent of all viewed Facebook ads.
31. See McKinsey and Company 2014.
32. See World Federation of Advertisers 2015.
33. See eMarketer 2014a.
34. See IPG Mediabrands 2014.
35. See U.S. marketing 2006; Dawson 2005.
36. See World Bank 2015.
37. See McKinsey and Company 2014.
38. See Borgie 2014.
39. See Census.gov 2017.
40. See Census.gov 2017.
41. See Statista 2017.
42. See Schiller 2014.
43. See McChesney 2013.
44. See Angwin 2010; Thielman 2015.
45. See Schiller 2014, 129.
46. See Foster and McChesney 2014; Mosco 2014.
47. See Fuchs 2012a, 141.
48. See Schiller 2014.
49. See Singer 2012.
50. See Acxiom.com 2015, para. 1.
51. See Zuboff 2019.
52. See Lohr 2009.
53. See Hubbard 2010.

54. See Gomes and Pavan 2017.
55. See eMarketer 2014b.
56. See Angwin 2010.
57. See Manyika et al. 2011.
58. See Mosco 2014.
59. See Office of Fair Trading 2010.
60. See Turow 2011.
61. See Reinartz 2002, 55; Ghose and Huang 2009; Obermiller, Arnesen, and Cohen 2012.
62. See Ghose and Huang 2009.
63. See Turow, Feldman, and Meltzer 2005, 10.
64. See Caillaud and De Nijs 2014.
65. See Shapiro and Varian 1999.
66. See Shiller 2013.
67. See Newman 2015.
68. See Reinartz 2002, 58.
69. See Hannak et al. 2014.
70. See Turow 2011.
71. See Turow 2011, 104.
72. See Elmaghraby and Keskinocak 2003; Ghose and Huang 2009; Obermiller, Arnesen, and Cohen 2012; Sahay 2007; Armstrong and Vickers 2001.
73. See Tanner 2014.
74. See Tuchman, Nair, and Gardete 2015, 6.
75. See Shiller 2013.
76. See Shiller 2013, 16.
77. On numerous occasions the author contacted several corporations that use PCS for interviews but none returned the emails or phone calls.
78. See Reinartz 2002.
79. See Fuchs 2015, 200.
80. See Turow 2011.
81. See Obermiller, Arnesen, and Cohen 2012.
82. See Hannak et al. 2014.
83. See Tanner 2014.
84. See Personalizing 2012.
85. See Homa 2015; Valentino-Devries, Singer-Vine, and Soltani 2012.
86. See Obermiller, Arnesen, and Cohen 2012; Shapiro and Varian 1999.
87. See International Air Transport Association 2015.
88. See Hannak et al 2014.
89. See Salinas 2013.
90. See Facefirst 2015, para. 1.
91. See Rocca 2015.
92. See Allevate 2013.
93. See Salinas 2013.
94. See Rubens 2014.
95. See Web-based targeted 2008, para. 1.
96. See Hill 2015.

97. See O'Connor 2014; Perlberg 2014.
98. See Tuchman, Nair, and Gardete 2015.
99. See Tuchman, Nair, and Gardete 2015.
100. See Taylor 2014.
101. See Saad-Fihlo 2003.
102. See Arvidsson 2011; Arvidsson and Colleoni 2012; Andrejevic 2011, Fuchs 2012b; Hesmondhalgh 2010.
103. See Fuchs 2010, 2012b; Andrejevic 2011, 2015.
104. See Fuchs 2012a.
105. See Fuchs 2015.
106. See Fuchs 2015.
107. See Newman 2014.
108. See Andrejevic 2015, 6.
109. It is important to note there is a difference between the value and price of a commodity (Fuchs 2012b). Value results from the production process (meaning working hours, prices of currency), while price is historically a measure of the circulation process, or selling of the commodities. The "transformation problem" can be addressed by converting the rate of surplus value into the rate of profit and, second, by converting the value of the commodity into the production price. The "transformation problem" and Marx's efforts to accurately price items are much debated and critiqued (Morishima 1973).
110. See Morishima 1973.
111. See Fuchs 2012b, 634.
112. See Andrejevic 2011, 283.
113. See Wright 1997.
114. See Burston, Dyer-Witheford, and Hearn 2010, 219.
115. See Zwick, Bonsu, and Darmody 2008.
116. See Zwick, Bonsu, and Darmody 2008.
117. See Armstrong and Vickers 2001; Ghose and Huang 2009; Obermiller, Arnesen, and Cohen 2012.
118. See Hirschman 2011.
119. See Valentino-Devries, Singer-Vine, and Soltani 2012.
120. See Newman 2014, 857.
121. See Tanner 2014.
122. Personal communication in 2015 via email.
123. See Lewis 2013, 59.
124. See Hannak et al. 2014.
125. See Obermiller, Arnesen, and Cohen 2012.
126. See Park 2009.
127. See Hannak et al. 2014; Klock 2002.
128. See Ellison and Fisher-Ellison 2009.
129. See Newman 2015.
130. See Ghose and Huang 2009.
131. See Andrejevic 2011, 281.
132. See Andrejevic 2015, 10.
133. See Hannak et al. 2014.

134. See McChesney 2013.
135. See Dunn 2017.
136. See Dunn 2017.
137. See Desjardins 2017.
138. See Newman 2014.
139. See Fottrell 2012.
140. See Reich 2015.
141. See McChesney 2013, 131.
142. See Investopedia 2017.
143. See Reich 2015.
144. See Bagdikian 1989; Borden 1942; Shiller 2013.
145. See Bagdikian 1989, 125.
146. See Norris 1989, 111.
147. See Wyatt 2013.
148. See Ritzer and Jurgenson 2010.
149. See Dyer-Witheford 1999.
150. See Schiller 2014.
151. See Negroponte 1996; Benkler 2006; Rezabakhsh et al. 2006.
152. See Goldstein 1994; Pavlik 1998.
153. See Browne 2015; Eubanks 2018; Noble 2018.
154. See Rosenblat et al. 2014.
155. See Bridle 2017; Eubanks 2018; O'Neil 2017; Srnicek 2017; Taplin 2017; Vaidhyanathan 2018.
156. See Turow 2011; Schiller 1996, 2014.
157. See Turow 2011.

Chapter 7

1. See Rauch 2018, 7.
2. See Jokinen, Malaska, and Kaivo-oja 1998.
3. See Mosco 2014.
4. See IPCC 2014b, 288.
5. See Chancel and Piketty 2015; OXFAM 2015.
6. See IPCC 2014b, 288.
7. See OXFAM 2015.
8. See Goodwin et al. n.d.
9. See Maxwell and Miller 2012a, 2012b.
10. See Williams 2004.
11. See Kuehr and Williams 2003.
12. See Christensen and Nilsson 2018.
13. See Baldé et al. 2017.
14. See Maxwell and Miller 2012b.
15. See Christensen and Nilsson 2018.
16. See Taylor 2014.
17. See Sterne 2007.
18. See Packard 1960.

19. See Lebow 1955.
20. See Park 2009.
21. See Hackett 2016.
22. See Grossman 2010.
23. See Leonard 2011.
24. See Zuboff 2019.
25. See Wicker and Karlsson 2017.
26. See Mosco 2014.
27. See Maxwell and Miller 2012a.
28. See Data Center Journal 2013.
29. See Glanz 2012.
30. See Greenpeace International 2010.
31. See Greenpeace 2019.
32. See Global video streaming 2020.
33. See Schiller 2014, 174.
34. See Wingfield 2016.
35. See Maxwell and Miller 2012a.
36. See Hackett 2016.
37. See Rauch 2018.
38. See Gartner 2007.
39. See IPCC 2014.
40. See U.S. Energy Information Administration 2016.
41. See Asara et al. 2015, n.p.
42. See InfluenceMap 2019.
43. According to economist James Kenneth Galbraith (2012), the majority of income growth leading up to the great recession of 2008 went to a small number of people in technology, military contracting, and finance sectors.
44. See Congressional Budget Office 2007; Schutz 2011.
45. See Carroll and Summer 1991.
46. See Information Observatory 2016.
47. See McChesney 2013.
48. See Rushe 2018.
49. See Kutsch 2015.
50. See Knight Frank Research 2016.
51. See Neate 2020.
52. See Milne 2015.
53. See Kollewe 2016.
54. See Tuttle 2013.
55. See Knight Frank Research 2016.
56. See Newman 2014; Stiglitz 2002.
57. See White House 2015, 6.
58. See Nichols and McChesney 2013; McChesney 2013.
59. See Piketty 2014.
60. See Schutz 2011, 5; Buchheit 2015.
61. See Saez and Zucman 2014.
62. See Wolff 2012.

63. See Collins and Hoxie 2015.
64. See McChesney 2013, 33.
65. See McKernan, Ratcliffe, and Cellini 2009.
66. See Buchheit 2015.
67. See National Coalition for the Homeless 2009.
68. See Piketty 2014.
69. See McChesney 2013, 14.
70. See Confessore, Cohen, and Yourish 2015.
71. See Garces and Rendall 2012.
72. See Nichols and McChesney 2013, 255–56.
73. See *Economist* 2020.
74. See Zuboff 2019.
75. See Rauch 2018, 7.
76. See Rauch 2018, 75.

Chapter 8

1. See https://regions20.org/2019/06/17/less-talk-more-action-at-3rd-r20-austrian-world-summit/

2. See Wilson 1995. Also the "US public receives a fairly uniform body of news, whether directly from the national organs, or indirectly through local ones" (Mazur and Lee 1993, 682). When asked where they get their news, respondents in a US survey indicated they received their news from television (57%), followed by newspapers (40%), followed by radio (36%), and last the internet (23%) (Pew Research Center for People and the Press 2006).

3. See Anderson 2011; Ryghaug, Sørensen, and Næss 2011; Stamm, Clark, and Reynolds-Eblacas 2000.

4. See Carvalho 2010; Stamm, Clark, and Reynolds-Eblacas 2000.
5. See Weber and Stern 2011.
6. See Happer and Philo 2016.
7. See Schuldt and Roh 2014.
8. See Entman 1993.
9. See Ytterstad 2015.
10. See Carmody 1995.
11. See McQuail 1977.
12. See Burgess 1990.
13. See Mazur and Lee 1993, 682.
14. See Boykoff and Rajan 2007.
15. See Boykoff and Roberts 2007.
16. See Mazur and Lee 1993.
17. See McCright 2006.
18. See Boykoff and Roberts 2007.
19. See McCright 2006.
20. See McChesney 2013.
21. See McChesney 2019.
22. See McIntryre 2019.

23. See Deuze, Elefante, and Steward 2010; Fuchs 2014; Gibson et al. 2015.
24. See Arnold 2019.
25. See "Whitening of the media" 2020.
26. See Poynter 2020.
27. See Gabbatt 2020.
28. See McChesney 2013.
29. See Boykoff 2013.
30. See Sparks 2018.
31. See Gibson et al. 2015.
32. See Hiles and Hinnant 2014.
33. See Dunwoody and Peters 1992.
34. See Hiles and Hinnant 2014.
35. See Bennett 2002; McComas and Shanahan 1999.
36. See Dunwoody and Peters 1992.
37. See Smith 2005.
38. See Hiles and Hinnant 2014.
39. See Good 2008.
40. See Tankard and Ryan 1974.
41. See Fahy and Nisbet 2011.
42. See Hackett 2016.
43. See Hiles and Hinnant 2014.
44. See Waterson 2019.
45. See Sparks 2018.
46. See Glenn Greenwald, Twitter.com, May 15, 2020, 8:26 am. https://twitter.com/ggreenwald/status/1261271794817929217
47. See Gibson et al. 2015. The purpose of this section is not to blame journalists for the discrepancies in climate change coverage. If anything, by now it should be obvious that they work in complex hierarchical organizations full of multiple pressures and power dynamics. They are but one piece of the puzzle.
48. See Moser 2010.
49. See McChesney 2013.
50. See Philo and Happer 2013. Not all studies found a decrease in climate change coverage. A study by Schmidt, Ivanova, and Schäfer (2013) found that climate change coverage increased in 27 countries around the world, which suggests it continues to be a globally relevant topic. However, many of the countries in the study, such as Australia, still contained carbon-intensive economies.
51. See Watts et al. 2018.
52. See Weathers and Kendall 2015.
53. It is important to highlight that the 2007 recession negatively impacted the news industry with substantial layoffs.
54. See Kalhoefer 2017.
55. See Kalhoefer 2017.
56. See Macdonald and Hymas 2019.
57. See Simon 2019.
58. See Watts et al. 2018.
59. See Boykoff and Rajan 2007.

60. See Good 2008.
61. See Watts et al. 2018.
62. See Boykoff 2008.
63. See Trumbo 1996.
64. See Takahashi et al. 2017.
65. See Takahashi et al. 2017.
66. See Boykoff and Farrell 2020.
67. See Boykoff and Farrell 2020, 131.
68. See Zehr 2000, 86.
69. See Boykoff and Boykoff 2007.
70. See Gibson et al. 2015.
71. See Brüggemann and Engesser 2014.
72. See Feldman, Maibach, Roser-Renouf, and Leiserowitz 2012.
73. See Feldman, Myers, Hmielowski, and Leiserowitz 2014.
74. See Feldman, Maibach, Roser-Renouf, and Leiserowitz 2012.
75. See Nuccitelli 2015.
76. See Boykoff and Farrell 2020.
77. See Ereaut and Segnit 2006.
78. See Gavin, Leonard-Milsom, and Montgomery 2011.
79. See Hiles and Hinnant 2014, 441.
80. See Harbinson, Mugara, and Chawla 2006.
81. See Schudson 1989.
82. See Gibson et al. 2015.
83. See Harbinson, Mugara, and Chawla 2006.
84. See Pope and Hertsgaard 2019.
85. See Nosty 2014.
86. See Opel 2006.
87. See Shanahan 2007.
88. See Gelbspan 2005.
89. See Anderson 2009, 170.
90. See Park, Wang, and Pinto 2015.
91. See Boykoff 2007, 485.
92. See Jackson 2019.
93. See Gibson et al. 2015.
94. See Gibson et al. 2015.
95. See Olausson 2009.
96. See Hackett 2016.
97. See Good 2008.
98. See Grant 2016.
99. See MacDonald 2018.
100. See Public Citizen 2018.
101. See Public Citizen 2019.
102. See Cooper 2019.
103. See Public Citizen 2019.
104. See Democracy Now! 2016.
105. See Johnson 2017.

106. See Pope and Hertsgaard 2019.
107. See Carvalho and Burgess 2005.
108. See Hackett 2016.
109. See Gavin, Leonard-Milsom, and Montgomery 2011; Henderson-Sellers 1998.
110. See Lozada, Kritz, and Mintu-Wimsatt 2017.
111. See Lozada, Kritz, and Mintu-Wimsatt 2017.

Chapter 9

1. See Aronoff 2019.
2. See Boykoff and Farrell 2020.
3. See Faber 2008.
4. See Faber 2008.
5. See Boussalis and Coan 2016.
6. See Goldenberg and Bengtsson 2015.
7. See Goldenberg and Bengtsson 2015.
8. See Boykoff and Boykoff 2004.
9. See Jurado and Maldonado 2014.
10. See Anderson 2009; Lewis and Boyce 2009.
11. See InfluenceMap 2019.
12. See "Kochland" 2019.
13. See Oreskes and Conway 2011.
14. See Shanahan 2007.
15. See Antilla 2005.
16. See Anderson 2009.
17. See Jacques, Dunlap, and Freeman 2008.
18. See Greitemeyer 2013.
19. See Arlt, Hoppe, and Wolling 2011.
20. See Miller-Gaither and Gaither 2016.
21. See Boykoff 2013.
22. See Carvalho 2010.
23. See Boykoff and Roberts 2007.
24. See Watson et al., 2016, 1.
25. See Nuccitelli 2018, n.p.
26. See Whitmarsh 2011.
27. If the media are interested in trying to connect again with a broader range of audiences, especially skeptical audiences, they may want to reconsider their present modes of operation. The media need to move away from wealthy elite audiences that advertisers so desire and try to connect with a broader economic spectrum of "consumers." In particular, the *New York Times* needs to reach out to working-class people and move away from the liberal elite, conservative war hawks, and liberal interventionists. This move may be challenging as the media prefer upper class groups, which are more lucrative. As Republican politicians continue their constant critique of the *New York Times*, many conservatives and climate skeptics will likely move away from reading the paper if they haven't already. Without reaching out to

new audiences, the effectiveness of the *New York Times* and other media outlets in terms of reaching deniers and skeptics should be questioned.

28. See Pew Research Center 2010.
29. See InfluenceMap 2019.
30. See Milman 2020.
31. See Lewis and Boyce 2009.
32. See Elgesem, Steskal, and Diakopoulos 2015.
33. See Tabuchi 2018.
34. See Matthews 2015.
35. See Elgesem, Steskal, and Diakopoulos 2015.
36. See Tabuchi 2018.
37. See Elgesem, Steskal, and Diakopoulos 2015; Koteyko, Jaspal, and Nerlich 2012.
38. See Weber and Stern 2011.
39. See Fottrell 2018.
40. See Boykoff and Roberts 2007; Weber and Stern 2011.
41. See Monmouth University Polling Institute 2018.
42. See Boykoff and Roberts 2007, 3.
43. See IPCC 2014.
44. See Matthews 2015.
45. See Boykoff and Smith 2010.
46. See Mazur and Lee 1993.
47. See Mazur and Lee 1993.
48. See Pidgeon 2012.
49. See Weathers and Kendall 2015.
50. See Brulle, Carmichael, and Jenkins 2012.
51. See Pew Research Center 2016.
52. See Lozada, Kritz, and Mintu-Wimsatt 2017.
53. See Leiserowitz et al. 2018.
54. See Leiserowitz et al. 2016; Leiserowitz et al. 2018.
55. See McCright and Dunlap 2011.
56. See Leiserowitz et al. 2018.
57. See Leiserowitz et al. 2016.
58. See Leiserowitz et al. 2016.
59. See Pew Research 2017.
60. See Pearson and Schuldt 2015.
61. See Leiserowitz et al. 2016.
62. See Van der Linden et al. 2015.
63. See Milman and Harvey 2019.
64. See Van der Linden et al. 2015.
65. See Hackett 2016.
66. See Hackett 2018, 102.
67. See Carvalho 2010.
68. See Block 2010.
69. See Shanahan 2007.
70. See Boykoff and Smith 2010.

71. See Hiles and Hinnant 2014.
72. See Good 2008.

Chapter 10

1. See Carvalho 2007; Jameson 1994.
2. See Baran and Sweezy 2013, 45.
3. See Foster and McChesney 2013.
4. See Mueller 1989.
5. See Foster and McChesney 2013. Even though some journalists have been able to "follow the money" regarding climate change denial information, companies and politicians with vested interests in maintaining the status quo continue efforts to undermine science and subvert journalism (Shanahan 2007).
6. See Splichal and Mance 2018.
7. See Foster 2017.
8. See Sparks 2018.
9. See Herman and Chomsky 2010.
10. See Herman and Chomsky 2010.
11. See Tuchman 1978.
12. See Ettema 2005; Chesebro and McMahan 2006; Kitch and Hume 2008.
13. See McCright and Dunlap 2003, 353.
14. See Gibson et al. 2015.
15. See Boykoff 2013.
16. See Hackett 2016.
17. See Baker 1994.
18. See Schutz 2011, 126.
19. See Hackett 2016, 12.
20. See Morgan 2015, 78.
21. See Morgan 2015.
22. See Schutz 2011, 127.
23. See Morgan 2015, 78.
24. See Good 2008, 239.
25. See Gilens and Hertzman 2000; Schoenfeld, Meier, and Griffin 1979.
26. See Anderson 2009; Takahashi 2011.
27. See Fuchs 2015, 326.
28. See Carvalho 2010.
29. See Carvalho 2010, 175.
30. This is a shame as the ability to cover environmental issues should not be taken for granted in the United States and other countries, as covering these topics in many other places in the world can either be illegal or attract a punitive action. Indeed, "coverage of environmental issues are governed by prevailing laws, and in some countries governments consider it subversive for the media to discuss issues like rainforest loss, bio-diversity or dam projects. Ironically, the economically advanced Asian countries in general have the most restrictive media while the poorer ones have a noisy and rambunctious free press" (Dixit 1994, 3).
31. See Schutz 2011, 127.

32. See Park and Wright 2007.
33. See Olausson 2010.
34. See Peters and Heinrichs 2008.
35. See Weingart, Engels, and Pansegrau 2000; Olausson 2009. The Swedish study examined dates from 2004 and 2005.
36. See Olausson 2009.
37. See Olausson 2009.
38. See Gibson et al. 2015.
39. The nonprofit *Guardian* newspaper displays some of the best mainstream media environmental reporting on global warming compared to most English-speaking media in the West. Its nonprofit status may help. However, not all of its sections earn that praise, as the foreign policy and political sections often appear to contain campaigns from powerful interests. This appears especially true when the paper writes about Russia, Syria, and the Ukraine. For example, when the Islamic State is fighting in Iraq, the *Guardian* often calls them terrorists, but when the same group or affiliated groups are fighting in Syria, they are called rebels.
40. See Ereaut and Segnit 2006.
41. See Post and Ramirez 2018.
42. See Carvalho 2005.
43. See Artz 2012, 540.
44. See Good 2008.
45. See Shanahan 2007.
46. See Emmons 2016.
47. See Emmons 2016.
48. See McChesney 2013.
49. See Reporters without Borders 2016.
50. See Oltermann 2016.
51. See Häussler et al. 2016.
52. See Weber and Stern 2011.
53. See Nielsen Company 2007.
54. See Arlt, Hoppe, and Wolling 2011.
55. See Brüggemann and Engesser 2014, 400.
56. See Gibson et al. 2015.
57. See Milman and Harvey 2019.
58. See Fuchs 2015, 332.
59. See Park et al. 2004.
60. See Perry 2015.
61. See Pew Research Center 2009.
62. See Pew Research Center 2013.
63. See Feldman and Swanson 2016.
64. See Reich 2019.
65. See Gibson et al. 2015.
66. See McCright 2006.
67. See Boykoff 2007.
68. See Roscho 1975.
69. See Van Dijk 2013.

70. See Nyhan 2014.
71. See Watson 2014.
72. While "skills" training is very important in journalism, it is unable to address the unfolding structural media crisis inflicting climate crisis coverage. For this reason, communication programs should also encourage and support scholars in their fields. Scholars, and critical scholars in particular, are the ones most likely to offer policy solutions to the larger, systemic challenges facing journalism and media in general.
73. See Althaus et al. 1996.
74. See Kinsella 2014.
75. See Hestres 2013.
76. See Arlt, Hoppe, and Wolling 2011, 60.
77. See Hackett 2016.
78. See Lewis and Boyce 2009, 10.
79. https://www.xrebellion.nyc/media-standards
80. For all of the conservative complaints about the media, conservative politicians are the least likely to support legislation to democratize the media by enabling a larger public media footprint. They know very well that a privatized, corporate media system is more likely to work in their favor than any public-oriented system. For this reason, proposals to create more independent, publicly owned, and publicly run media will not come from Republican legislators. Argentina represents a good example of how conservative political parties reject media systems with a significant percentage of public and nonprofit media. Several hundred citizen groups were able to successfully push legislation democratizing the highly concentrated commercial Argentine media system (which was a relic of the Dirty War dictatorship) under the administration of Néstor Kirchner. After the elections placed the conservative opposition into the presidency, one of the first items on the agenda was to recommercialize the media, which would convert the newly emboldened public sector back to a few mega-media conglomerates.
81. See McChesney 2013; Park, Wang, and Pinto 2015.
82. See Pickard 2019.
83. See Pickard 2019, 165.
84. The question is who will pay for this? Eliminating the US tax breaks for the rich while increasing the tax rates on corporations and the wealthy can begin to transfer money back to the poor and middle classes, as well as funding a larger public media system.

Conclusion

1. See Pickard 2019.
2. See Holden and Gambino 2019.
3. See Perry 2019.
4. See Laville 2019; Waterson 2020.
5. See Hestres 2020.
6. See Karlin 2019.
7. See Hertsgaard 2019.

8. See Krishnaswamy 2019.
9. See White 2019.
10. See Krishnaswamy 2019.

References

Introduction

Almiron, Núria. 2020. Rethinking the ethical challenge in climate lobbying: A discussion about denial. In *Climate change denial and public relations: Strategic communication and interest groups in climate inaction*, edited by Núria Almiron and Jordi Xifra, 9–25. New York: Routledge.

Boykoff, M. T., and J. M. Boykoff. 2007. Climate change and journalistic norms: A case-study of US mass-media coverage. *Geoforum* 38 (6): 1190–1204.

Chancel, L., and T. Piketty. 2015. Carbon and inequality: From Kyoto to Paris. Trends in the global inequality of carbon emissions (1998–2013) and prospects for an equitable adaptation fund. Paris School of Economics. http://piketty.pse.ens.fr/files/ChancelPiketty2015SummaryEN.pdf

Corbett, J. B., and J. L. Durfee. 2004. Testing public (un)certainty of science media representations of global warming. *Science Communication* 26 (2): 129–51.

Foster, J. 2013. The epochal crisis. *Monthly Review* 65 (5): 1–12.

Hales, S., S. Kovats, S. Lloyd, and D. Campbell-Lendrum, eds. 2014. Quantitative risk assessment of the effects of climate change on selected causes of death, 2030s and 2050s. Geneva: World Health Organization.

Hansen, J. 2009. *Storms of my grandchildren*. New York: Bloomsbury.

IPCC. 2014. *Mitigation of climate change*. Contribution of Working Group III to the Fifth Assessment Report of the Intergovernmental Panel on Climate Change. Edited by O. Edenhofer, R. Pichs-Madruga, Y. Sokona, E. Farahani, S. Kadner, K. Seyboth, A. Adler, I. Baum, S. Brunner, and P. Eickemeier, et al. Cambridge: Cambridge University Press.

OXFAM. 2015. Extreme carbon inequality: Why the Paris climate deal must put the poorest, lowest emitting and most vulnerable people first. Oxford Committee for Famine Relief (OXFAM), December 2. https://www.oxfam.org/sites/www.oxfam.org/files/file_attachments/mb-extreme-carbon-inequality-021215-en.pdf

Revkin, A. 2008. Global heating, atmosphere cancer, pollution death: What's in a

name? *Dot Earth*, NYTimes.com blog, February 18. http://dotearth.blogs.nytimes.com/2008/02/18
Sayers, Dorothy. 1947. *Creed or Chaos?* London: Methuen.
Stempel, G., and B. Westley. 1989. *Research methods in mass communication*, 2nd ed. Englewood Cliffs, NJ: Prentice-Hall.
Storch, H. von, and W. Krauss. 2005. Culture contributes to perceptions of climate change: A comparison between the United States and Germany reveals insights about why journalists in each country report about this issue in different ways. *Nieman Reports*, Winter 2005. http://niemanreports.org/articles/culture-contributes-to-perceptions-of-climate-change/

1. The Last Generation to Stop the Disaster

Adams, Rod. 2013. Climate scientists propose radical solutions—including nuclear energy. Atomicinsights.com. https://atomicinsights.com/climate-scientists-propose-radical-solutions-including-nuclear-energy/
Almiron, Núria. 2020. Rethinking the ethical challenge in climate lobbying: A discussion about denial. In *Climate change denial and public relations: Strategic communication and interest groups in climate inaction*, edited by Núria Almiron and Jordi Xifra, 9–25. New York: Routledge.
American Association for the Advancement of Science Climate Science Panel. 2014. What we know: The reality, risks, and response to climate change. American Association for the Advancement of Science.
Anderson, K. 2012. Climate change going beyond dangerous—brutal numbers and tenuous hope. *Development Dialogue* 61 (September): 16–40.
Anderson, Kevin. 2013. Avoiding dangerous climate change demands de-growth strategies from the wealthier nations. November 25. http://kevinanderson.info/blog
Baran, P., and P. Sweezy. 2013. Theses on advertising. *Monthly Review* 65 (3): 34–42.
Boussalis, C., and T. G. Coan. 2016. Text-mining the signals of climate change doubt. *Global Environmental Change* 36: 89–100.
Boykoff, M. T., and J. T. Roberts. 2007. Media coverage of climate change: Current trends, strengths, weaknesses. *Human Development Report 2007/2008*, 1–53. New York: United Nations Development Programme.
Burton, I., S. Huq, B. Lim, O. Pilifosova, and E. L. Schipper. 2002. From impacts assessment to adaptation priorities: The shaping of adaptation policy. *Climate Policy* 2 (2–3): 145–59.
Carrington, Damian. 2018a. David Attenborough: Collapse of civilization is on the horizon. *Guardian.com*, December 3. https://www.theguardian.com/environment/2018/dec/03/david-attenborough-collapse-civilisation-on-horizon-un-climate-summit
Carrington, Damian. 2018b. "We are the last generation that can stop climate change"—UN summit. *Guardian.com*, December 2. https://www.theguardian.com/environment/2018/dec/03/we-are-last-generation-that-can-stop-climate-change-un-summit
Carrington, Damian. 2019. Climate emergency: World "may have crossed tipping

points." *Guardian.com*, November 27. https://www.theguardian.com/environment/2019/nov/27/climate-emergency-world-may-have-crossed-tipping-points

Clark, P., et al. 2016. Consequences of twenty-first-century policy for multi-millennial climate and sea-level change. *Nature Climate Change* 6: 360–69.

Climate Change Performance Index 2019. 2019. Germanwatch.org. https://www.climate-change-performance-index.org

Corbett, J. B., and J. L. Durfee. 2004. Testing public (un)certainty of science media representations of global warming. *Science Communication* 26 (2): 129–51.

D'Alisa, Giacomo, Federico Demaria, and Giorgos Kallis. 2014. *Degrowth: A vocabulary for a new era*. New York: Routledge.

Daly, H. 1993. Sustainable growth: An impossibility theorem in valuing the earth. In *Valuing the earth: Economics, ecology, ethics*, edited by H. Daly and K. Townsend, 276–78. Cambridge, MA: MIT Press.

Democracy Now. 2018. Climate scientist: As U.N. warns of global catastrophe, we need a "Marshall Plan" for climate change. *Democracynow.org*, October 9. https://www.democracynow.org/2018/10/9/climate_scientist_as_un_warns_of

Foster, J. 2015. Marxism and ecology: Common fronts of a great transition. *Monthly Review* 67 (7): 1–13.

Foster, John Bellamy, and Brett Clark. 2015. Crossing the river of fire: The liberal attack on Naomi Klein and *This Changes Everything*. *MonthlyReview.org*, February. https://monthlyreview.org/2015/02/01/crossing-the-river-of-fire/

Goldenberg, Suzanne. 2013. UN's 2C target will fail to avoid a climate disaster, scientists warn: Global warming limit agreed is too late and dangerous as a 1C rise in temperature will trigger catastrophic events, study says. *Guardian.com*, December 3. https://www.theguardian.com/environment/2013/dec/03/un-2c-global-warming-climate-change

Hales, S., S. Kovats, S. Lloyd, and D. Campbell-Lendrum, eds. 2014. Quantitative risk assessment of the effects of climate change on selected causes of death, 2030s and 2050s. Geneva: World Health Organization.

Hansen, J. 2009. *Storms of my grandchildren*. New York: Bloomsbury.

InterAcademy Partnership, 2018. Opportunities for future research and innovation on food and nutrition security and agriculture: The InterAcademy Partnership's Global Perspective. http://www.interacademies.org/48898/Opportunities-for-future-research-and-innovation-on-food-and-nutrition-security-and-agriculture-The-InterAcademy-Partnerships-global-perspective

IPCC. 2014a. Intergovernmental Panel on Climate Change: Fifth assessment report (AR5). http://www.ipcc.ch/activities/activities.shtml

IPCC. 2014b. *Mitigation of climate change*. Contribution of Working Group III to the Fifth Assessment Report of the Intergovernmental Panel on Climate Change. Edited by O. Edenhofer, R. Pichs-Madruga, Y. Sokona, E. Farahani, S. Kadner, K. Seyboth, A. Adler, I. Baum, S. Brunner, and P. Eickemeier, et al. Cambridge: Cambridge University Press.

IPCC. 2018. Summary for Policymakers. In *Global warming of 1.5°C. An IPCC Special Report on the impacts of global warming of 1.5°C above pre-industrial levels and related global greenhouse gas emission pathways, in the context of strengthening the global response to the threat of climate change, sustainable development, and efforts to eradicate poverty*. Edited by V. Masson-Delmotte, P. Zhai, H. O. Pörtner, D. Roberts, J. Skea, P. R. Shukla, A. Pirani, W. Moufouma-Okia, C. Péan, and R. Pidcock, et al. Geneva: World Meteorological Organization.

Jacobson, M., and M. Delucchi. 2009. A path to sustainable energy by 2030. *Scientific American* 301 (November): 58–65. https://doi.org/10.1038/scientificamerican1109-58

Jhally, S. 1997. *Advertising and the end of the world: The movie*. PDF transcript. Media Education Foundation. https://www.mediaed.org/assets/products/101/transcript_101.pdf

Kallis, G. 2011. In defense of degrowth. *Ecological Economics* 70 (5): 873–80.

Karl, T. R., and K. E. Trenberth. 2003. Modern global climate change. *Science* 302: 1719–23.

Latouche, S. 2010. Degrowth. *Journal of Cleaner Production* 18: 519–22.

Leary, Elly, and Anne Lewis. 2015. Interview with Bill Gallegos. *Monthyreview.org*, October. https://monthlyreview.org/2015/10/01/interview-with-bill-gallegos/

Lewis, J. 2013. *Beyond consumer capitalism: Media and the limits of imagination*. Boston: Polity Press.

Lewis, J., and T. Boyce. 2009. Climate change and the media: The scale of the challenge. In *Climate change and the media*, edited by J. Lewis and T. Boyce, 3–16. New York: Peter Lang.

Martinez-Alier, J., U. Pascual, F. Vivien, and E. Zaccai. 2010. Sustainable de-growth: Mapping the context, criticisms and future prospects of an emergent paradigm. *Ecological Economics* 69: 1741–47.

Miller, Brandon, and Jay Croft. 2018. Planet has only until 2030 to stem catastrophic climate change, experts warn. *Cnn.com*, October 8. https://www.cnn.com/2018/10/07/world/climate-change-new-ipcc-report-wxc/index.html

Milman, Oliver. 2018a. World "nowhere near on track" to avoid warming beyond 1.5C target. *Guardian.com*, September 27. https://www.theguardian.com/environment/2018/sep/26/global-warming-climate-change-targets-un-report

Milman, Oliver. 2018b. Ex-NASA scientist: 30 years on, world is failing "miserably" to address climate change. *Guardian.com*, June 19. https://www.theguardian.com/environment/2018/jun/19/james-hansen-nasa-scientist-climate-change-warning

Oil Change International. 2016. The sky's limit: Why the Paris climate goals require a managed decline of fossil fuel production. Oil Change International, September. http://priceofoil.org/content/uploads/2016/09/OCI_the_skys_limit_2016_FINAL_2.pdf

Perin, C. 2005. *Shouldering risks: The culture of control in the nuclear power industry*. Princeton: Princeton University Press.

Pielke, R. A. 1998. Rethinking the role of adaptation in climate policy. *Global Environmental Change* 8 (2): 159–70.

Revkin, A. 2008. Global heating, atmosphere cancer, pollution death. What's in a name? *Dot Earth*, NYTimes.com blog, February 18. http://dotearth.blogs.nytimes.com/2008/02/18

Scheffers, Brett, Luc De Meester, Tom C. L. Bridge, Ary A. Hoffmann, John M. Pandolfi, Richard T. Corlett, Stuart H. M. Butchart, Paul Pearce-Kelly, Kit M.

Kovacs, David Dudgeon, et al. 2016. The broad footprint of climate change from genes to biomes to people. *Science* 354 (6313). https://doi.org/10.1126/science.aaf7671

Shapiro, S. 2014. Poor people, poor planet: The psychology of how we harm and heal humanity and earth. In *Toward a socially responsible psychology for a global era*, edited by E. Mustakova-Possardt, M. Lyubansky, M. Basseches, and J. Oxenberg, 231–54. New York: Springer. https://doi.org/10.1007/978-1-4614-7391-6

Stanton, E., and F. Ackerman. 2011. Nuclear power: Not the solution to climate change. *Monthly Review*, May 9. http://mrzine.monthlyreview.org/2011/sa090511.html

Steffen, Will, Johan Rockström, Katherine Richardson, Timothy M. Lenton, Carl Folke, Diana Liverman, Colin P. Summerhayes, Anthony D. Barnosky, Sarah E. Cornell, and Michel Crucifix et al. 2018. Trajectories of the Earth system in the Anthropocene. *Proceedings of the National Academy of Sciences* 115 (33): 8252–59. https://doi.org/10.1073/pnas.1810141115

Storch, H. V., and W. Krauss. 2005. Culture contributes to perceptions of climate change: A comparison between the United States and Germany reveals insights about why journalists in each country report about this issue in different ways. *Neuman Report*, Winter 2005. http://niemanreports.org/articles/culture-contributes-to-perceptions-of-climate-change/

Sweeney, S. 2014. Working toward energy democracy. In *State of the world 2014: Governing for sustainability*, edited by the Worldwatch Institute, 215–27. Washington, DC: Island Press.

Union of Concerned Scientists. 2012. U.S. nuclear power one year after Fukushima. September 23. https://www.ucsusa.org/resources/us-nuclear-power-safety-one-year-after-fukushima

United Nations Environment Programme. 2018. Emissions gap report. United Nations. http://wedocs.unep.org/bitstream/handle/20.500.11822/26896/EGR-KEYMESSAGES_2018.pdf?sequence=1&isAllowed=y

U.S. Energy Information Administration. 2014. *Total energy: Annual energy review (primary energy consumption by source and sector, 2012)*. http://www.eia.gov/totalenergy/data/monthly/pdf/flow/primary_energy.pdf

United States Global Change Research Program. 2018. *Impacts, risks, and adaptation in the United States: Fourth national climate assessment, volume II*. Edited by D. R. Reidmiller, C. W. Avery, D. R. Easterling, K. E. Kunkel, K. L. M. Lewis, T. K. Maycock, and B. C. Stewart. 1,515 pages. Washington, DC: U.S. Global Change Research Program. https://doi.org/10.7930/NCA4.2018

United States Nuclear Regulatory Commission. 2002. *Radioactive waste: Production, storage, disposal*. http://www.nrc.gov/reading-rm/doc-collections/nuregs/brochures/br0216/r2/br0216r2.pdf

Warf, J. C., and S. C. Plotkin. 1995. *Disposal of high-level nuclear waste*. Nuclear Age Peace Foundation. http://dwb4.unl.edu/Chem/CHEM869Z/CHEM869ZLinks/www.napf.org/resources/gs23.html

Watson, R., C. Carraro, P. Canziani, N. Nakicenovic, J. McCarthy, J. Goldemberg, and L. Hisas. 2016. The truth about climate change. Universal Ecological

Fund. https://drive.google.com/drive/folders/0B7HRGKIJWol-OEZfNDFjR HUwM2c

Watts, N., M. Amann, N. Arnell, S. Ayeb-Karlsson, K. Belesova, H. Berry, and D. Campbell-Lendrum. 2018. The 2018 report of the Lancet Countdown on health and climate change: Shaping the health of nations for centuries to come. *The Lancet* 392 (10163): 2479–2514.

Wilkins, L. 1993. Between facts and values: Print media coverage of the greenhouse effect, 1987–1990. *Public Understanding of Science* 2 (1): 71–84.

2. US Political Will to Address the Climate Crisis

American Lung Association. 2019. Key findings: The state of the air 2019. www.lung.org. https://www.lung.org/our-initiatives/healthy-air/sota/key-findings/

Aronoff, Kate. 2018. Young leftist candidates are breathing new radicalism into stale climate politics. *Intercept.com*, July 5. https://theintercept.com/2018/07/05/climate-change-left-politics-policy/

Aronoff, Kate. 2019. Exxon Mobil is funding centrist Democratic think tank, disclosures reveal. *Intercept.com*, September 6. https://theintercept.com/2019/09/06/exxon-mobil-progressive-policy-institute-climate/

Associated Press. 2018a. Big oil asks government to protect it from climate change. *Nola.com*, August 22. https://www.nola.com/environment/index.ssf/2018/08/big_oil_asks_government_to_pro.html

Associated Press. 2018b. Senior EPA officials collaborated with climate change denial group, emails show. *Guardian.com*, May 25. https://www.theguardian.com/environment/2018/may/26/senior-epa-officials-collaborated-with-climate-change-denial-group-emails-show

BBC. 2017. Chomsky: Republican Party "most dangerous organization on Earth." *BBC.com*, May 11. https://www.bbc.com/news/av/world-us-canada-39879374/chomsky-republican-party-most-dangerous-organisation-on-earth

Boykoff, M. T., and S. R. Rajan. 2007. Signals and noise. *EMBO Reports* 8 (3): 207–11.

Brown, Alleen. 2019. Trump administration asks Congress to make disrupting pipeline construction a crime punishable by 20 years in prison. *Intercept.com*, June 5. https://theintercept.com/2019/06/05/pipeline-protests-proposed-legislation-phmsa-alec/

Brulle, R. J. 2000. *Agency, democracy, and nature: The US environmental movement from a critical theory perspective*. Cambridge, MA: MIT Press.

Carrington, D. 2015. Fossil fuels subsidized by $10m a minute, says IMF. *Guardian.com*, May 18. https://www.theguardian.com/environment/2015/may/18/fossil-fuel-companies-getting-10m-a-minute-in-subsidies-says-imf

Carvalho, A. 2005. Representing the politics of the greenhouse effect: Discursive strategies in the British media. *Critical Discourse Studies* 2 (1): 1–29.

Cutler, D., and F. Dominici. 2018. A breath of bad air: Cost of the Trump environmental agenda may lead to 80,000 extra deaths per decade. *JAMA* 319 (22): 2261–62. https://doi.org/10.1001/jama.2018.7351

Democracy Now! 2016. Trump's war on science: Exxon CEO expected to head

State, while "enemies list" alarms Energy Dept. *Democracynow.org*, December 12. https://www.democracynow.org/2016/12/12/trumps_war_on_science_exxon_ceo

Democracy Now! 2019. Headlines. *Democracynow.org*, July 19. https://www.democracynow.org/2019/7/19/headlines

DNC lifts ban on fossil fuel company donations. 2018. *Democracynow.org*, August 14. https://www.democracynow.org/2018/8/14/headlines/dnc_lifts_ban_on_fossil_fuel_company_donations

Dodd, Vikram, and Jamie Grierson. 2020. Greenpeace included with neo-nazis on UK counter-terror list. *Guardian.com*, January 17. https://www.theguardian.com/uk-news/2020/jan/17/greenpeace-included-with-neo-nazis-on-uk-counter-terror-list

EPA indefinitely suspends enforcement of environmental laws. 2020. *Democracynow.org*, March 27. https://www.democracynow.org/2020/3/27/headlines

EPA to introduce new counting model that downplays projected death toll from Trump coal plan. 2019. *Democracynow.org*, May 21. https://www.democracynow.org/2019/5/21/headlines

Ewen, S. 2001. *Captains of consciousness: Advertising and the social roots of consumer culture*. 25th anniversary ed. New York: Basic Books.

Federman, Adam. 2020. Revealed: US listed climate activist group as "extremists" alongside mass killers. *Guardian.com*, January 13. https://www.theguardian.com/environment/2020/jan/13/us-listed-climate-activist-group-extremists

Fountain, Henry. 2016. Trump's climate contrarian. *New York Times*, November 11. https://www.nytimes.com/2016/11/12/science/myron-ebell-trump-epa.html

Friedman, Lisa. 2019. White House tried to stop climate science testimony, documents show. *New York Times*, June 8. https://www.nytimes.com/2019/06/08/climate/rod-schoonover-testimony.html

Goldenberg, S., and H. Bengtsson. 2015. Secretive donors gave US climate denial groups $125m over three years. *Guardian.com*, June 9. https://www.theguardian.com/environment/2015/jun/09/secretive-donors-gave-us-climate-denial-groups-125m-over-three-years

Grandoni, Dino. 2018. The Energy 202: Trump puts former BP oil spill lawyer in charge of environmental law enforcement. *Washington Post*, October 12. https://www.washingtonpost.com/news/powerpost/paloma/the-energy-202/2018/10/12/the-energy-202-trump-puts-former-bp-oil-spill-lawyer-in-charge-of-environmental-law-enforcement/5bbf88411b326b7c8a8d1948/?noredirect=on&utm_term=.69936df3dd86

Guardian Staff. 2019. Trump administration expected to roll back methane regulations—report. *Guardian.com*, August 29. https://www.theguardian.com/environment/2019/aug/29/trump-administration-roll-back-methane-regulations

Hackett, R. A. 2016. Dallas Smythe Award keynote lecture. *Democratic Communiqué* 27: 1–14.

Harish, Alon. 2012. New law in North Carolina bans latest scientific predictions of sea-level rise. *ABCnews.go.com*, August 2. ABCnew.go.com/us/north-carolina-bans-latest-science-rising-sea-level

Holden, Emily. 2019a. David Bernhardt: Trump selects former oil lobbyist to oversee Interior Department. *Guardian.com*, February 4. https://www.theguardian.com/us-news/2019/feb/04/david-bernhardt-trump-oil-lobbyist-interior-department

Holden, Emily. 2019b. White House climate change panel to include man who touted emissions. *Guardian.com*, February 20. https://www.theguardian.com/environment/2019/feb/20/white-house-climate-change-national-security-panel

Holden, Emily. 2020a. 10 US oil refineries exceeding limits for cancer-causing benzene, report finds. *Guardian.com*, February 6. https://www.theguardian.com/business/2020/feb/06/us-oil-refineries-exceeding-limits-for-cancer-causing-benzene-report-finds

Holden, Emily. 2020b. Oil and gas industry rewards US lawmakers who oppose environmental protections—study. *Guardian.com*, February 24. https://www.theguardian.com/environment/2020/feb/24/oil-gas-industry-us-lawmakers-campaign-donations-analysis

Holden, Emily. 2020c. Trump to roll back Obama-era clean car rules in huge blow to climate fight. *Guardian.com*, March 31. https://www.theguardian.com/environment/2020/mar/31/trump-epa-obama-clean-car-rules-climate-change

Horgan, John. 2018. Noam Chomsky calls Trump and Republican allies "criminally insane." *Scientificamerican.com*, November 3. https://blogs.scientificamerican.com/cross-check/noam-chomsky-calls-trump-and-republican-allies-criminally-insane/

Interior Dept. official peppers reports with lies about climate change. 2020. *Democracynow.org*, March 3. https://www.democracynow.org/2020/3/3/headlines

Jacques, P. J., R. E. Dunlap, and M. Freeman. 2008. The organization of denial: Conservative think tanks and environmental skepticism. *Environmental Politics* 17 (3): 349–85.

Jamieson, Dale. 2014. *Reason in a dark time: Why the struggle against climate change failed—and what it means for our future*. New York: Oxford University Press.

Kahan, D. M., D. Braman, J. Gastil, P. Slovic, and C. K. Mertz. 2007. Culture and identity-protective cognition: Explaining the white-male effect in risk perception. *Journal of Empirical Legal Studies* 4 (3): 465–505.

Korten, Tristram. 2015. In Florida, officials ban term "climate change." *Miami Herald*, March 8. https://www.miamiherald.com/news/state/florida/article12983720.html

Labor Network. 2015. *The clean energy future: Protecting the climate, creating jobs and saving money*. Labor4sustainability.org. https://www.labor4sustainability.org/wp-content/uploads/2015/10/cleanenergy_10212015_main.pdf

Lacy, Akela. 2019. After climate forum, Biden heads to a fundraiser co-hosted by a fossil fuel executive. *Intercept.com*, September 4. https://theintercept.com/2019/09/04/after-climate-forum-biden-heads-to-a-fundraiser-co-hosted-by-a-fossil-fuel-executive/

LaMotte, Sandee. 2020. Air quality in US dramatically worse than in prior years, says new "State of the Air" report. *CNN.com*, April 21. https://edition.cnn.com/2020/04/21/health/air-quality-2020-report-wellness/index.html

Lartey, Jamiles. 2017. Another US agency deletes references to climate change on

government website. *Guardian.com*, August 23. https://www.theguardian.com/us-news/2017/aug/23/us-government-agency-climate-change-references-removed-again
Lewis, J., S. Inthorn, and K. Wahl-Jorgensen. 2005. *Citizens or consumers: The media and the decline of political participation.* Maidenhead, UK: Open University Press.
Lewis, Renee. 2015. Analysts: Fossil fuel firms could waste $2.2 trillion on unneeded projects. *Aljazeera.com*, November 25. http://america.aljazeera.com/articles/2015/11/25/fossil-fuel-firms-risk-destroying-investor-returns.html
Long, Heather. 2016. American CEOs are thrilled with Trump. *CNN.com*, December 6. https://money.cnn.com/2016/12/06/news/economy/donald-trump-ceos/
Marsh, Rene. 2017a. EPA pulls scientists out of climate change conference talk. *CNN.com*, October 24. https://www.cnn.com/2017/10/23/politics/epa-rhode-island/index.html
Marsh, Rene. 2017b. Trump administration dismisses climate change advisory panel. *CNN.com*, August 21. https://www.cnn.com/2017/08/21/politics/white-house-climate-change-committee-dismantled/index.html
Matthews, P. 2015. Why are people skeptical about climate change? Some insights from blog comments. *Environmental Communication* 9 (2): 153–68.
McCright, A. M. 2006. Dealing with climate change contrarians. In *Creating a climate for change*, edited by S. Moser and L. Dilling, 200–212. Cambridge: Cambridge University Press.
McCright, A. M., and R. E. Dunlap. 2000. Challenging global warming as a social problem: An analysis of the conservative movement's counter-claims. *Social Problems* 47 (4): 499–522.
McCright, A. M., and R. E. Dunlap. 2003. Defeating Kyoto: The conservative movement's impact on US climate change policy. *Social Problems* 50 (3): 348–73.
McCright, A. M., and R. E. Dunlap. 2011. The politicization of climate change and polarization in the American public's views of global warming, 2001–2010. *Sociological Quarterly* 52 (2): 155–94.
McKnight, Laura. 2019. Trump to roll back drilling rules prompted by the Deepwater Horizon disaster: Report. *Nola.com*, May 1. https://www.nola.com/news/environment/article_34334b1d-0784-512e-9b2d-b1987a22b4db.html
Milman, Oliver. 2015a. James Hansen, father of climate change awareness, calls Paris talks "a fraud." *Guardian.com*, December 12. https://www.theguardian.com/environment/2015/dec/12/james-hansen-climate-change-paris-talks-fraud
Milman, Oliver. 2015b. Top scientists accuse House panel of harassing climate researchers. *Guardian.com*, November 25. https://www.theguardian.com/science/2015/nov/25/leading-scientists-accuse-house-panel-of-harassing-climate-researchers
Milman, Oliver. 2016a. Donald Trump picks climate change sceptic Scott Pruitt to lead EPA. *Guardian.com*, December 8. https://www.theguardian.com/us-news/2016/dec/07/trump-scott-pruitt-environmental-protection-agency
Milman, Oliver. 2016b. Donald Trump presidency a "disaster for the planet," warn climate scientists. *Guardian.com*, November 11. https://www.theguardian.com

/environment/2016/nov/11/trump-presidency-a-disaster-for-the-planet-climate-change

Milman, Oliver. 2016c. Trump's transition: Sceptics guide every agency dealing with climate change. *Guardian.com*, December 12. https://www.theguardian.com/us-news/2016/dec/12/donald-trump-environment-climate-change-skeptics

Milman, Oliver. 2017a. EPA removes half of scientific board, seeking industry-aligned replacements. *Guardian.com*, May 8. https://www.theguardian.com/us-news/2017/may/08/epa-board-scientific-scott-pruitt-climate-change

Milman, Oliver. 2017b. "No shame": How the Trump administration granted big oil's wishlist. *Guardian.com*, December 12. https://www.theguardian.com/us-news/2017/dec/12/big-oil-lobby-get-what-it-wants-epa-trump-pruitt

Milman, Oliver. 2017c. US federal department is censoring use of term "climate change," emails reveal. *Guardian.com*, August 7. https://www.theguardian.com/environment/2017/aug/07/usda-climate-change-language-censorship-emails

Milman, Oliver. 2018. Ex-NASA scientist: 30 years on, world is failing "miserably" to address climate change. *Guardian.com*, June 19. https://www.theguardian.com/environment/2018/jun/19/james-hansen-nasa-scientist-climate-change-warning

Nuccitelli, Dana. 2018. Pruitt promised polluters EPA will value their profits over American lives. *Guardian.com*, April 23. https://www.theguardian.com/environment/climate-consensus-97-per-cent/2018/apr/23/pruitt-promised-polluters-epa-will-value-their-profits-over-american-lives

O'Neil, Luke. 2019. US Energy Department rebrands fossil fuels as "molecules of freedom." *Guardian.com*, May 29. https://www.theguardian.com/business/2019/may/29/energy-department-molecules-freedom-fossil-fuel-rebranding

OXFAM. 2015. Extreme carbon inequality: Why the Paris climate deal must put the poorest, lowest emitting and most vulnerable people first. Oxford Committee for Famine Relief (OXFAM), December 2. https://www.oxfam.org/sites/www.oxfam.org/files/file_attachments/mb-extreme-carbon-inequality-021215-en.pdf

Park, D., V. Wang, and J. Pinto. 2015. Beyond nuclear disaster and risk: Comparison of post-Fukushima news in U.S. and German press. *Communication, Culture and Critique*, July 29. https://doi.org/10.1111/cccr.12119

Prasad, S., J. Burke, M. Slezak, and O. Milman. 2016. Obama's dirty secret: The fossil fuel projects in the US littered around the world. *Guardian.com*, December 1. https://www.theguardian.com/environment/2016/dec/01/obama-fossil-fuels-us-export-import-bank-energy-projects

Report: All major oil companies knew of climate change by 1970s. 2015. *Democracynow.org*, December 24. http://www.democracynow.org/2015/12/24/headlines/report_all_major_oil_companies_knew_of_climate_change_by_1970s

Reports: EPA won't set limits on harmful chemicals in drinking water. 2019. *Democracynow.org*, February 1. https://www.democracynow.org/2019/2/1/headlines/wsj_after_venezuela_us_to_target_cuba_in_effort_to_reshape_latin_america

Rosenberg, Paul. 2019. Climate debate-gate: Democratic Party chokes on its own greenhouse gas—but the fight's not over. *Salon.com*, June 22. https://www.salon

.com/2019/06/22/climate-debate-gate-democratic-party-doubles-down-on-stupid-but-the-fights-not-over

Sneath, Sara. 2020. Air pollution could make coronavirus more severe for some Louisianans. *Nola.com*, April 7. https://www.nola.com/news/coronavirus/article_1f3514be-781f-11ea-abd8-53f40536c7f6.html

Soraghan, Mike. 2018. Trump team reluctant to pursue polluters—report. *Eenews.net*, November 16. https://www.eenews.net/greenwire/stories/1060106485

Splichal, S. 2009. "New" media, "old" theories: Does the (national) public melt into the air of global governance? *European Journal of Communication* 24 (4): 391–405.

Symons, Jeremy. 2019. The DNC's climate problems run deep. *Thehill.com*, June 18. https://thehill.com/opinion/energy-environment/449177-the-dncs-climate-problems-run-deep

Trump admin escalates multipronged attack on climate science. 2019. *Democracynow.org*, May 28. https://www.democracynow.org/2019/5/28/headlines

Venezuela in crisis: As U.S. pushes regime change, fear grows of civil war and famine. 2019. *Democracynow.org*, February 19. https://www.democracynow.org/2019/2/19/venezuela_in_crisis_as_us_pushes

Verburg, Steven. 2016. DNR wipes cause of climate change from website. *Madison.com*, December 30. http://host.madison.com/wsj/news/local/environment/dnr-wipes-cause-of-climate-change-from-website/article_c8785fac-5e55-5c18-97fc-7369bed14dc8.html

Ward, Alex. 2019. Andrew McCabe claims Trump wanted war in Venezuela because "they have all that oil." *Vox.com*, February 20. https://www.vox.com/world/2019/2/20/18233394/mccabe-trump-venezuela-war-oil-lawrence

WHO says meeting Paris climate goal would save millions of lives. 2018. *Democracynow.org*, December 6. https://www.democracynow.org/2018/12/6/headlines

Wilson, Jason. 2019. How Republicans killed Oregon's climate crisis bill—by fleeing the state. *Guardian.com*, June 29. https://www.theguardian.com/us-news/2019/jun/28/how-republican-senators-killed-oregons-climate-crisis-bill

Zuboff, S. 2019. *The age of surveillance capitalism: The fight for a human future at the new frontier of power.* New York: PublicAffairs.

3. Advertising, Consumerism, Industrialism, and Ideology

Andreucci, Diego, and Terrence McDonough. 2014. Capitalism. In *Degrowth: A vocabulary for a new era*, edited by G. D'Alisa, F. Demaria, and Giorgos Kallis, 86–89. New York: Routledge.

Asara, V., I. Otero, F. Demaria, and E. Corbera. 2015. Socially sustainable degrowth as a social-ecological transformation: Repoliticizing sustainability. *Sustainability Science* 10 (3): 375–84.

Bagdikian, Ben. 1989. The high cost of free lunches. In *Advertising in society: Classic and contemporary readings on advertising's role in society*, edited by Roxanne Hovland and Gary Wilcox, 119–33. Lincolnwood, IL: NTC Business Books.

Baran, P., and P. Sweezy. 2013a. The quality of monopoly capitalist society: Culture and communications. *Monthly Review* 65 (3): 43–64.

Baran, P., and P. Sweezy. 2013b. Theses on advertising. *Monthly Review* 65 (3): 34–42.
Beckwith, B. P. 1949. *The economic theory of a socialist economy*. Stanford: Stanford University Press.
Berry, M. 2013. *The affluent society revisited*. Oxford: Oxford University Press.
Brulle, R., and L. E. Young. 2007. Advertising, individual consumption levels, and the natural environment, 1900–2000. *Sociological Inquiry* 77: 522–42.
Carey, James. 1989. Advertising: An institutional approach. In *Advertising in society: Classic and contemporary readings on advertising's role in society*, edited by Roxanne Hovland and Gary Wilcox, 11–26. Lincolnwood, IL: NTC Business Books.
Comor, E. 2008. *Consumption and the globalization project: International hegemony and the annihilation of time*. New York: Palgrave Macmillan.
Cook, D. 2011. *Adorno on nature*. Durham, NC: Acumen.
Corrado, C., C. Hulten, and D. Sichel. 2009. Intangible capital and US economic growth. *Review of Income and Wealth* 55 (3): 661–85.
Dahlen, M., and S. Rosengren. 2016. If advertising won't die, what will it be? Toward a working definition of advertising. *Journal of Advertising* 45 (3): 334–45.
Dominick, J. 2009. *The dynamics of mass communication*. 2nd ed. New York: McGraw-Hill.
Driver, J., and G. Foxall. 1984. *Advertising, policy, and practice*. New York: St. Martin's Press.
Dyer, G. 1982. *Advertising as communication*. New York: Methuen.
Eagleton, T. 1991. *Ideology: An introduction*. London: Verso.
Ellul, J. 1965. *Propaganda*. New York: Knopf.
eMarketer. 2014a. Total US ad spending to see largest increase since 2004: Mobile advertising leads growth; will surpass radio, magazines, and newspapers this year. *eMarketer*, July 2. http://www.emarketer.com/Article/Total-US-Ad-Spending-See-Largest-Increase-Since-2004/1010982
eMarketer. 2014b. Total US retail sales top $4.5 trillion in 2013, outpace GDP growth. *eMarketer*, April 10. http://www.emarketer.com/Article/Total-US-Retail-Sales-Top-3645-Trillion-2013-Outpace-GDP-Growth/1010756
Ewen, S. 2001. *Captains of consciousness: Advertising and the social roots of consumer culture*. 25th anniversary ed. New York: Basic Books.
Foster, J. 2015. Marxism and ecology: Common fronts of a great transition. *Monthly Review* 67 (7): 1–13.
Foster, J. B., and R. W. McChesney. 2013. The cultural apparatus of monopoly capital: An introduction. *Monthly Review* 65 (3): 1–33.
Fuchs, C. 2015. *Culture and economy in the age of social media*. New York: Routledge.
Fuentes, C. 2014. Green materialities: Marketing and the socio-material construction of green products. *Business Strategy and the Environment* 23: 105–16. https://doi.org/10.1002/bse.1768
Gerth, K. 2013. Compromising with consumerism in socialist China: Transnational flows and internal tensions in "socialist advertising." *Past & Present* 218 (8): 203–32.
Ghotge, Sanjeev. 2018. Climate change and Marx in the twenty-first century, part II. *Capitalism Nature Socialism* 29 (3): 11–20. https://doi.org/10.1080/10455752.2018.1474565

Goodwin, N., J. Nelson, A. Ackerman, and T. Weisskopf. N.d. Consumption and the consumer society. Global Development and Environment Institute, Tufts University. http://www.ase.tufts.edu/gdae/education_materials/modules/Consumption_and_the_Consumer_Society.pdf

Gorz, A. 1994. *Capitalism, socialism, ecology*. London: Verso.

Gunderson, R. 2015. Environmental sociology and the Frankfurt School 1: Reason and capital. *Environmental Sociology* 1 (3): 224–35.

Hamilton, C. 2009. Consumerism, self-creation, and prospects for a new ecological consciousness. *Journal of Cleaner Production* 18: 571–75.

Handley, L. 2016. Global advertising spending to slow in 2017, while 2016 sales reached nearly $500bn: Research. *CNBC.com*, December 5. http://www.cnbc.com/2016/12/05/global-ad-spend-to-slow-in-2017-while-2016-sales-were-nearly-500bn.html

Hanson, P. 1974. *Advertising and socialism: The nature and extent of consumer advertising in the Soviet Union, Poland, Hungary and Yugoslavia*. London: Springer.

Heilbroner, R., and L. Thurow. 1998. *Economics explained: Everything you need to know about how the economy works and where it's going*. New York: Touchstone.

Held, D. 1980. *Introduction to critical theory: Horkheimer to Habermas*. Berkeley: University of California Press.

Holm, N. 2017. Art for fun and profit: The political aesthetics of advertising. In *Explorations in critical studies of advertising*, edited by J. Hamilton, R. Bodle, and E. Korin, 97–109. New York: Routledge.

Horkheimer, M., and T. W. Adorno. 2002. *Dialectic of enlightenment*. Stanford: Stanford University Press.

IPCC. 2014. *Mitigation of climate change*. Contribution of Working Group III to the Fifth Assessment Report of the Intergovernmental Panel on Climate Change. Edited by O. Edenhofer, R. Pichs-Madruga, Y. Sokona, E. Farahani, S. Kadner, K. Seyboth, A. Adler, I. Baum, S. Brunner, and P. Eickemeier, et al. Cambridge: Cambridge University Press. https://www.ipcc.ch/report/ar5/syr/

IPG Mediabrands. 2014. MAGNA GLOBAL forecasts global advertising revenues to grow by +4.8% to $536 billion in 2015. December 8. http://news.ipgmediabrands.com/article_display.cfm?article_id=1666

Jhally, S. 1997. *Advertising and the end of the world: The movie*. PDF transcript. Media Education Foundation. https://www.mediaed.org/assets/products/101/transcript_101.pdf

Jowett, G., and V. O'Donnell. 2006. *Propaganda and persuasion*. Thousand Oaks, CA: Sage.

Kallis, Giorgos. 2017. Socialism without growth. *Capitalism Nature Socialism* 30 (2): 189–206.

Kallis, G., V. Kostakis, S. Lange, B. Muraca, S. Paulson, and M. Schmelzer. 2018. Research on degrowth. *Annual Review of Environment and Resources* 43: 291–316.

Kanner, A., and M. Gomes. 1995. The all consuming self. In *Ecopsychology: Restoring the earth—healing the mind*, edited by T. Roszak, M. E. Gomes, and A. D. Kanner. San Francisco: Sierra Club Books.

Kilbourne, W. E. 1995. Green advertising: Salvation or oxymoron? *Journal of Advertising* 24 (2): 7–20.

Klein, Naomi. 2018. Capitalism killed our climate momentum, not "human nature." *Intercept.com*, August 3. https://theintercept.com/2018/08/03/climate-change-new-york-times-magazine/

Kornai, J. 1992. *The socialist system: The political economy of communism*. London: Oxford University Press.

Lewis, J., S. Inthorn, and K. Wahl-Jorgensen. 2005. *Citizens or consumers: The media and the decline of political participation*. Maidenhead, UK: Open University Press.

Mad men are watching you. 2011. *The Economist*, May 5. http://www.economist.com/node/18651104

Marx, K. 1973. *Grundrisse*. Harmondsworth, UK: Penguin.

McKinsey and Company. 2014. Global media report: Global industry overview. http://www.mckinsey.com/industries/media-and-entertainment/our-insights/global-media-report-2014

Media Education Foundation. N.d. Media education foundation study guide: Toxic sludge is good for you. http://www.mediaed.org/discussion-guides/Toxic-Sludge-is-Good-for-You-Discussion-Guide.pdf

Morgan, T. 2015. Growing ourselves to death? Economic and ecological crises, the growth of waste, and the role of the media and cultural industries. *Human Geography* 8 (1): 68–81.

Norris, V. 1989. Advertising history—according to the textbooks. In *Advertising in society: Classic and contemporary readings on advertising's role in society*, edited by Roxanne Hovland and Gary Wilcox, 99–118. Lincolnwood, IL: NTC Business Books.

Northrup, C. C. 2003. *The American economy*. Santa Barbara, CA: ABC-CLIO.

O'Connor, James. 1989. Political economy of ecology of socialism and capitalism. *Capitalism Nature Socialism* 1 (3): 93–107. https://doi.org/10.1080/10455758909358386

Packard, V. 1960. *The waste makers*. New York: David McKay Company.

Pearlin, L., and M. Rosenberg. 1956. Propaganda techniques in institutional advertising. In *Public opinion and propaganda*, edited by D. Katz, D. Cartwright, S. Eldersveld, and A. McClung-Lee. New York: Dryden Press.

Pepper, D. 2002. *Eco-socialism: From deep ecology to social justice*. New York: Routledge.

Pineault, E. 2018. From provocation to challenge: Degrowth, capitalism, and the prospect of "socialism without growth": A commentary on Giorgios Kallis. *Capitalism Nature Socialism* 30 (2): 251–66.

Pollay, Richard. 1989. The distorted mirror: Reflections on the unintended consequences of advertising. In *Advertising in society: Classic and contemporary readings on advertising's role in society*, edited by Roxanne Hovland and Gary Wilcox, 437–76. Lincolnwood, IL: NTC Business Books.

Press Association. 2016. "New era of climate change reality" as emissions hit symbolic threshold. *Guardian.com*, October 24. https://www.theguardian.com/environment/2016/oct/24/new-era-of-climate-change-reality-as-emissions-hit-symbolic-threshold

Richards, J. I., and C. M. Curran. 2002. Oracles on "advertising": Searching for a definition. *Journal of Advertising* 31 (2): 62–77.

Rotzoll, K., J. Haefner, and C. Sandage. 1976. *Advertising in contemporary society: Perspectives toward understanding*. Columbus: Grid.
Saxton, J. 2017. Why your customers' attention is the scarcest resource in 2017. American Marketing Association. https://www.ama.org/partners/content/Pages/why-customers-attention-scarcest-resources-2017.aspx
Scatamburlo-D'Annibale, V. 2010. Beyond the culture jam. In *Critical pedagogies of consumption: Living and learning in the shadow of the "Shopocalypse"*, edited by J. Sandlin and P. McLaren, 224–35. New York: Routledge.
Schor, J. 1999. *The overspent American*. New York: Harper Perennial.
Schudson, Michael. 1989. Historical roots of consumer culture. In *Advertising in society: Classic and contemporary readings on advertising's role in society*, edited by Roxanne Hovland and Gary Wilcox, 42–72. Lincolnwood, IL: NTC Business Books.
Schutz, E. 2011. *Inequality and power: The economics of class*. New York: Routledge.
Simon, J. 1970. *Issues in the economics of advertising*. Urbana: University of Illinois Press.
Storey, J. 2017. *Theories of consumption*. New York: Routledge.
Vergara-Camus, L. 2019. Capitalism, democracy, and the degrowth horizon. *Capitalism Nature Socialism* 30 (2): 217–33.
Williams, C. 2010. *Ecology and socialism*. Chicago: Haymarket Books.
World Federation of Advertisers. 2015. What is the WFA? http://www.wfanet.org/en/about-wfa
Zukin, S. 2004. *Point of purchase: How shopping changed American culture*. New York: Routledge.

4. Advertising and Its Interdependence with the Origins of the Climate Crisis

Ackerberg, D. 2001. Empirically distinguishing informative and prestige effects of advertising. *RAND Journal of Economics* 32 (2): 316–33.
Ackley, G. 1961. *Macroeconomic theory*. New York: Macmillan.
Agger, Ben. 1987. *Western Marxism: An introduction: Classical and contemporary sources*. Santa Monica, CA: Goodyear Publishing.
Alperovitz, G. 2014. Growth for growth's sake will kill us all: Moral and ecological truths are challenging economic doctrines. *Al Jazeera*, April 11. http://www.garalperovitz.com/2014/06/growth-growths-sake-will-kill-us-moral-ecological-truths-challenging-economic-doctrines/
Antilla, L. 2005. Climate of scepticism: US newspaper coverage of the science of climate change. *Global Environmental Change* 15 (4): 338–52.
Asara, Viviana, Iago Otero, Federico Demaria, and Esteve Corbera. 2015. Socially sustainable degrowth as a social-ecological transformation: Repoliticizing sustainability. *Sustainability Science* 10 (3): 375–84.
Assadourian, Erik. 2010. The rise and fall of consumer cultures. In *Transforming cultures: From consumerism to sustainability*, edited by E. Assodourian, 3–20. New York: W. W. Norton.
Baran, P., and P. Sweezy. 2013. Theses on advertising. *Monthly Review* 65 (3): 34–42.
Bauman, Z. 2007. *Consuming life*. Cambridge: Polity.

Becker, G., and K. Murphy. 1993. A simple theory of advertising as good or bad. *Quarterly Journal of Economics* 108 (4): 941–64.
Berry, M. 2013. *The affluent society revisited.* Oxford: Oxford University Press.
Berry, T. 1995. The viable human. In *Deep ecology for the 21st century*, edited by G. Sessions, 8–18. Boston: Shambhala Publications.
Boykoff, M. T., and J. M. Boykoff. 2007. Climate change and journalistic norms: A case-study of US mass-media coverage. *Geoforum* 38 (6): 1190–1204.
Brems, H. 1951. *Product equilibrium under monopolistic competition.* Cambridge, MA: Harvard University Press.
Brower, M., and W. Leon. 1999. *The consumer's guide to effective environmental choices: Practical advice from the Union of Concerned Scientists.* New York: Three Rivers Press.
Brown, L. R. 2001. *Eco-economy: Building an economy for the earth.* New York: W. W. Norton.
Brulle, R., and L. E. Young. 2007. Advertising, individual consumption levels, and the natural environment, 1900–2000. *Sociological Inquiry* 77: 522–42.
Carey, J. 1960. Advertising: An institutional approach. In *The role of advertising: A book of readings*, edited by C. Sandage and V. Fryburger, 3–17. Homewood, IL: Richard D. Irwin.
Chancel, L., and T. Piketty. 2015. Carbon and inequality: From Kyoto to Paris. Trends in the global inequality of carbon emissions (1998–2013) and prospects for an equitable adaptation fund. Paris School of Economics. http://piketty.pse.ens.fr/files/ChancelPiketty2015SummaryEN.pdf
Comor, E. 2008. *Consumption and the globalization project: International hegemony and the annihilation of time.* New York: Palgrave Macmillan.
Congressional Budget Office. 2007. Trade-offs in allocating allowances for CO2 emissions. Economic and Budget Issue Brief, 1–8. http://cbo.gov
Craig, R. T. 2018. For a practical discipline. *Journal of Communication* 68 (2): 289–97.
Daly, H. 1993. Sustainable growth: An impossibility theorem in valuing the earth. In *Valuing the Earth: Economics, ecology, ethics*, edited by H. Daly and K. Townsend, 267–74. Cambridge, MA: MIT Press.
DeFleur, M., and S. Ball-Rokeach. 1975. *Theories of mass communication*, 3rd ed. New York: David McKay.
Democracynow. 2018. Climate scientist: As U.N. warns of global catastrophe, we need a "Marshall Plan" for climate change. *Democracynow.org*, October 9. https://www.democracynow.org/2018/10/9/climate_scientist_as_un_warns_of
Dongling, C. 1999. *World consumption economics.* Singapore: World Scientific Publishing.
Donohew, L., and P. Plamgreen. 1989. Theory construction. In *Research methods in mass communication*, 2nd ed., edited by G. Stempel III and B. Westley, 30–48. Englewood Cliffs, NJ: Prentice-Hall.
Dyer, G. 1982. *Advertising as communication.* New York: Methuen.
Ewen, S. 2001. *Captains of consciousness: Advertising and the social roots of consumer culture.* 25th anniversary ed. New York: Basic Books.
Faraone, R. 2011. Economy, ideology, and advertising. In *The handbook of political*

economy of communications, edited by J. Wasco, G. Murdock, and H. Sousa, 187–205. Oxford: Blackwell.

Foster, J. B. 2013. The epochal crisis. *Monthly Review* 65 (5): 1–12.

Foster, J. B. 2017. *Trump in the White House: Tragedy and farce*. New York: Monthly Review Press.

Foster, J. B., B. Clark, and R. York. 2011. *The ecological rift: Capitalism's war on the earth*. New York: Monthly Review Press.

Frank, C. 2009. The bankruptcy of capitalist solutions to the climate crisis. *Capitalism Nature Socialism* 20 (20): 32–34.

Fuchs, C., and J. Linchuan Qiu. 2018. Ferments in the field: Introductory reflections on the past, present and future of communication studies. *Journal of Communication* 68 (2): 219–32.

Galbraith, J. 1969. *The affluent society*. Boston: Houghton Mifflin.

Gardner, G., and P. Stern. 2002. *Environmental problems and human behavior*. Boston: Pearson Custom.

Gardner, G., E. Assadourian, and R. Sarin. 2004. The state of consumption today. In *State of the world 2004*, edited by L. Starke. New York: W. W. Norton.

Ge, M., J. Friedrich, and T. Damassa. 2014. 6 graphs explain the world's top 10 emitters. World Resources Institute, November 25. http://www.wri.org/blog/2014/11/6-graphs-explain-world%E2%80%99s-top-10-emitters

Gerth, K. 2013. Compromising with consumerism in socialist China: Transnational flows and internal tensions in "socialist advertising." *Past & Present* 218 (8): 203–32.

Goodman, D., and M. Cohen. 2004. *Consumer culture*. Santa Barbara, CA: ABC-CLIO.

Goodwin, N., J. Nelson, A. Ackerman, and T. Weisskopf. N.d. Consumption and the consumer society. Global Development and Environment Institute, Tufts University. http://www.ase.tufts.edu/gdae/education_materials/modules/consumption_and_the_consumer_society.pdf

Greer, S. 1969. *The logic of social inquiry*. Chicago: Aldine Publishing.

Hackett, R. A. 2016. Dallas Smythe Award keynote lecture. *Democratic Communiqué* 27: 1–14.

Hamilton, C. 2009. Consumerism, self-creation and prospects for a new ecological consciousness. *Journal of Cleaner Production* 18 (6): 571–75.

Hardy, J. 2014. *Critical political economy of the media: An introduction*. New York: Routledge.

Harvey, David. 2014. *Seventeen contradictions and the end of capitalism*. London: Profile Books.

Heede, R. 2013. *Carbon majors: Accounting for carbon and methane emissions 1854–2010 methods & results report*. Climate Justice Programme, Greenpeace International and Climate Mitigation Services. http://www.climateaccountability.org/pdf/MRR%209.1%20Apr14R.pdf

Hirschman, D. 2011. On the road to riches: Those under 35 with $100K household income. *Advertising Age*, May 22. http://adage.com/article/adagestat/ll-rich-35-100k-household-income/227671/

Horowitz, D. 1994. *Vance Packard and American social criticism*. Chapel Hill: University of North Carolina Press.

InfluenceMap. 2019. March. Big oil's real agenda on climate change. https://influencemap.org/report/How-Big-Oil-Continues-to-Oppose-the-Paris-Agreement-38212275958aa21196dae3b76220bddc

IPCC. 2014. *Mitigation of climate change*. Contribution of Working Group III to the Fifth Assessment Report of the Intergovernmental Panel on Climate Change. Edited by O. Edenhofer, R. Pichs-Madruga, Y. Sokona, E. Farahani, S. Kadner, K. Seyboth, A. Adler, I. Baum, S. Brunner, and P. Eickemeier et al. Cambridge: Cambridge University Press. https://www.ipcc.ch/report/ar5/syr/

Jackson, T. 2010. *Prosperity without growth? The transition to a sustainable economy*. London: Sustainable Development Commission.

Jhally, S. 1997. *Advertising and the end of the world: The movie*. PDF transcript. Media Education Foundation. https://www.mediaed.org/assets/products/101/transcript_101.pdf

Jurado, D. G., and C. J. Maldonado. 2014. Advertising and climate change, relationships, and divergences. *Interactions: Studies in Communication & Culture* 5 (1): 107–25.

Just, N., and M. Puppis. 2018. Moving beyond self-castigation: Let's reinvigorate communication policy research now! *Journal of Communication* 68 (2): 327–36.

Kallis, Giorgos. 2017. Socialism without growth. *Capitalism Nature Socialism* 30 (2): 189–206. https://doi.org/10.1080/10455752.2017.1386695

Kallis, Giorgos, V. Kostakis, S. Lange, B. Muraca, S. Paulson, and M. Schmelzer. 2018. Research on degrowth. *Annual Review of Environment and Resources* 43: 291–316.

Kaplan, A. 1964. *The conduct of inquiry*. New York: Harper and Row.

Kerlinger, F. 1986. *Foundations in behavioral research*. New York: Holt, Rinehart and Winston.

Kindleberger, C. P. 1958. *International economics*. Rev. ed. Homewood, IL: R. D. Irwin.

Klein, N. 2014. *This changes everything: Capitalism versus the climate*. New York: Simon and Schuster.

Kotler, P., G. Armstrong, L. Brown, and S. Adam. 2006. *Marketing*, 7th ed. Melbourne, Victoria: Pearson Education Australia/Prentice Hall.

Kuznets, S. 1973. Modern economic growth: Findings and reflections. *American Economic Review* 63 (3): 247–58.

Magdoff, F. 2011. Ecological civilization. *Monthly Review* 62 (8): 1–25.

Magdoff, F. 2014. An ecologically sound and socially just economy. *Monthly Review* 66 (4): 23–34.

Magdoff, F., and J. B. Foster. 2011. *What every environmentalist needs to know about capitalism: A citizen's guide to capitalism and the environment*. New York: Monthly Review Press.

Maniates, M. 2002. Individualization: Plant a tree, buy a bike, save the world? In *Confronting consumption*, edited by T. Princen, M. Maniates, and K. Conca. Cambridge, MA: MIT Press.

Marx, K. 1973. *Grundrisse*. Harmondsworth, UK: Penguin.

McChesney, R. W. 2004. *The problem of the media*. New York: Monthly Review Press.

McDonald, C. 2015. How many Earths do we need? *BBC News*, June 6. http://www.bbc.com/news/magazine-33133712

Morgan, T. 2015. Growing ourselves to death? Economic and ecological crises, the growth of waste, and the role of the media and cultural industries. *Human Geography* 8 (1): 68–81.

O'Connor, J. 1998. *Natural causes: Essays in ecological Marxism.* New York: Guilford Press.

O'Connor, J. 1999. A prolegomenon to an ecological Marxism: Thoughts on the materialist conception of history. *Capitalism Nature Socialism* 10 (2): 77–106.

Odum, H. 2007. *Environment, power, and society.* New York: Columbia University Press.

Olausson, U. 2009. Global warming—global responsibility? Media frames of collective action and scientific certainty. *Public Understanding of Science* 18 (4): 421–36.

OXFAM. 2015. World's richest 10% produce half of carbon emissions while poorest 3.5 billion account for just a tenth. Oxfam.org, December 12. https://www.oxfam.org/en/pressroom/pressreleases/2015-12-02/worlds-richest-10-produce-half-carbon-emissions-while-poorest-35

Packard, V. 1960. *The waste makers.* New York: David McKay Company.

Pineault, Eric. 2017. From provocation to challenge: Degrowth, capitalism and the prospect of "socialism without growth," a commentary on Giorgos Kallis. *Capitalism Nature and Socialism*, April 12. https://doi.org/10.1080/10455752.2018.1457064

Polanyi, Karl. 1967. *The great transformation.* Boston: Beacon Press.

Ryazanov, V.T. 2015. Development of the theory of capitalist crisis: Political economy of traditions and modernity. *Экономика регион* 2: 41–50.

Reuters. 2019. The Pentagon emits more greenhouse gases than Portugal, study finds. *Guardian.com*, June 12. https://www.theguardian.com/environment/2019/jun/12/pentagon-greenhouse-gas-emissions-portugal

Reynolds, P. 1971. *A primer in theory construction.* New York: Macmillan.

Rockström, J., W. Steffen, K. Noone, Å Persson, F. S. Chapin, E. F. Lambin and A. J. Foley. 2009. A safe operating space for humanity. *Nature* 461: 472–75. https://doi.org/10.1038/461472a

Sato, M. 2012. Embodied carbon in trade: A survey of the empirical literature. Grantham Research Institute on Climate Change and the Environment. Working paper no. 77. http://cccep.ac.uk

Schumpeter, J. A. 1934. *The theory of economic development.* Cambridge, MA: Harvard University Press.

Schor, J. 1998. *The overspent American.* New York: Harper Perennial.

Schutz, E. 2011. *Inequality and power: The economics of class.* New York: Routledge.

Shah, Anup. 2014. Consumption and consumerism. Global Issues. www.globalissues.org/issue/235/consumption-and-consumerism

Shanahan, M. 2007. *Talking about a revolution: Climate change and the media.* London: International Institute for Environment and Development.

Shapiro, S. 2014. Poor people, poor planet: The psychology of how we harm and heal humanity and Earth. In *Toward a socially responsible psychology for a global era*, edited by E. Mustakova-Possardt, M. Lyubansky, M. Basseches, and J. Oxenberg, 231–54. New York: Springer.

Simon, J. 1970. *Issues in the economics of advertising.* Urbana: University of Illinois Press.
Sterne, J. 2007. Out with the trash: On the future of new media. In *Residual media*, edited by C. R. Acland, 16–31. Minneapolis: University of Minnesota Press.
Storey, J. 2017. *Theories of consumption.* New York: Routledge.
Sukhdev, P. 2012. Sustainability: The corporate climate overhaul. *Nature* 486 (7401): 27–28.
Suzuki, D. 1997. *The sacred balance: Rediscovering our role in nature.* Vancouver, British Columbia: Greystone Books.
Tichenor, P., and D. McLeod. 1989. The logic of social and behavioral science. In *Research methods in mass communication, 2nd ed.*, edited by G. Stempel III and B. H. Westley, 10–29. Englewood Cliffs, NJ: Prentice-Hall.
Tuchman, A., H. Nair, and P. Gardete. 2015. Complementarities in consumption and the consumer demand of advertising. Stanford University. http://web.stanford.edu/~tuchman/TuchmanNairGardete_ComplementaritiesAdvertisingConsumption.pdf
US Energy Information Administration. 2016. EIA projects 48% increase in world energy consumption by 2040. www.eia.gov/todayinenergy/detail.cfm?id=26212
Vergara-Camus, Leandro. 2019. Capitalism, democracy, and the degrowth horizon. *Capitalism Nature Socialism* 30 (2): 217–33. https://doi.org/10.1080/10455752.2017.1344868
Wackernagel, M., and W. Rees. 1996. *Our ecological footprint: Reducing human impact on human health.* Gabriola Island, British Columbia: New Society Publishers.
Watson, R., C. Carraro, P. Canziani, N. Nakicenovic, J. McCarthy, J. Goldemberg, and L. Hisas. 2016. The truth about climate change. Universal Ecological Fund. https://drive.google.com/drive/folders/0B7HRGKIJWol-OEZfNDFjRHUwM2c
Worldcentric. N.d. Social and economic justice. Worldcentric.org. http://www.worldcentric.org/conscious-living/social-and-economic-injustice
Yang, D. 1966. *Advertising and the economy.* New York: Mimeo.

5. Conceptualizing a Mitigative Model of Advertising

Apte, Uday M., Uday S. Karmarkar, and Hiranya K. Nath. 2007. Information services in the US economy: Value, jobs and management implications. http://www.ifm.eng.cam.ac.uk/ssme/references/Karmarkar_ref2_cambridgessme07.pdf
Artz, L. 2017. Commentary: Transnational capitalism and the new political economy of transnational media. *Political Economy of Communication* 5 (2): 95–101.
Assadourian, Erik. 2010. The rise and fall of consumer cultures. In *Transforming cultures: From consumerism to sustainability*, edited by Erik Assodourian, 3–20. New York: W. W. Norton.
Baran, P., and P. Sweezy. 2013. Theses on advertising. *Monthly Review* 65 (3): 34–42.
Brower, M., and W. Leon. 1999. *The consumer's guide to effective environmental choices: Practical advice from the Union of Concerned Scientists.* New York: Three Rivers Press.

Brown, L. R. 2001. *Eco-economy: Building an economy for the earth*. New York: W. W. Norton.

Budinsky, J., and S. Bryant. 2013. "It's not easy being green": The greenwashing of environmental discourses in advertising. *Canadian Journal of Communication* 38: 207–26.

Carey, James. 1960. Advertising: An institutional approach. In *The role of advertising: A book of readings*, edited by C. Sandage and V. Fryburger, 3–17. Homewood, IL: Richard D. Irwin.

Carrington, Damian. 2018. "We are the last generation that can stop climate change"—UN summit. *Guardian.com*, December 2. https://www.theguardian.com/environment/2018/dec/03/we-are-last-generation-that-can-stop-climate-change-un-summit

CDP. 2016. Press release: Major multinationals at forefront of drive to price carbon and meet climate targets but many companies still unprepared. September 19. https://www.cdp.net/es/articles/media/press-release-major-multinationals-at-forefront-of-drive-to-price-carbon-and-meet-climate-targets-but-many-companies-still-unprepared

Chancel, L., and T. Piketty. 2015. Carbon and inequality: From Kyoto to Paris. Trends in the global inequality of carbon emissions (1998–2013) and prospects for an equitable adaptation fund. Paris School of Economics. http://piketty.pse.ens.fr/files/ChancelPiketty2015SummaryEN.pdf

Comor, E. 2008. *Consumption and the globalization project: International hegemony and the annihilation of time*. New York: Palgrave Macmillan.

Craig, R. T. 2018. For a practical discipline. *Journal of Communication* 68 (2): 289–97.

Daly, H. 1993. Sustainable growth: An impossibility theorem in valuing the earth. In *Valuing the earth: Economics, ecology, ethics*, edited by H. Daly and K. Townsend, 276–78. Cambridge, MA: MIT Press.

Dangelico, R. M., and P. Pontrandolfo. 2015. Being "green and competitive": The impact of environmental actions and collaborations on firm performance. *Business Strategy and the Environment* 24 (6): 413–30.

Deeg, K., E. Lyon, A. Leiserowitz, E. Maibach, and J. Marlon. 2019. Who is changing their mind about global warming and why? Yale University and George Mason University. New Haven: Yale Program on Climate Change Communication.

Demaria, Federico, F. Schneider, F. Sekulova, and J. Martinez-Alier. 2013. What is degrowth? From an activist slogan to a social movement. *Environmental Values* 22 (2): 191–215.

Dominick, J. 2009. *The dynamics of mass communication*. 2nd ed. New York: McGraw-Hill.

Driver, John, and Gordon Foxall. 1984. *Advertising policy and practice*. New York: St. Martin's Press.

eMarketer. 2014, April 10. Total US retail sales top $4.5 trillion in 2013, outpace GDP growth. *eMarketer*. http://www.emarketer.com/Article/Total-US-Retail-Sales-Top-3645-Trillion-2013-Outpace-GDP-Growth/1010756

Foster, J. B. 2013. The epochal crisis. *Monthly Review* 65 (5): 1–12.

Fuchs, C. 2015. *Culture and economy in the age of social media*. New York: Routledge.
Fuchs, C., and J. Linchuan Qiu. 2018. Ferments in the field: Introductory reflections on the past, present and future of communication studies. *Journal of Communication* 68 (2): 219–32.
Fuentes, C. 2014. Green materialities: Marketing and the socio-material construction of green products. *Business Strategy and the Environment* 23: 105–16. https://doi.org/10.1002/bse.1768
Galbraith, John. 1969. *The affluent society*. Rev. ed. 1958. Reprint. Boston: Houghton Mifflin Company.
Gartner. 2007. Gartner estimates ICT industry accounts for 2 percent of global CO2 emissions. Gartner.com, April 26. http://www.gartner.com/newsroom/id/503867
Halperin, Alex. 2018. Up in smoke: New California law could send $350m worth of cannabis to incinerator. *Guardian.com*, June 29. https://www.theguardian.com/society/2018/jun/29/california-cannabis-laws-product-destroyed
Hamilton, C. 2009. Consumerism, self-creation and prospects for a new ecological consciousness. *Journal of Cleaner Production* 18: 571–75.
Handley, L. 2016. Global advertising spending to slow in 2017, while 2016 sales reached nearly $500bn: Research. *CNBC.com*, December 5. http://www.cnbc.com/2016/12/05/global-ad-spend-to-slow-in-2017-while-2016-sales-were-nearly-500bn.html
Hansen, A. 2010. *Environment, media and communication*. New York: Routledge.
Hansen, J. 2009. *Storms of my grandchildren*. New York: Bloomsbury.
Happer, C., and G. Philo. 2016. New approaches to understanding the role of the news media in the formation of public attitudes and behaviours on climate change. *European Journal of Communication* 31 (2): 136–51.
Hardy, J. 2014. *Critical political economy of the media: An introduction*. New York: Routledge.
Harris, Alex. 2018. Why Miami is the first to stop on a campaign to ask polluters to pay for climate change. *MiamiHerald.com*, March 9. https://www.miamiherald.com/news/local/environment/article204217279.html
Hartmann, P., V. Apaolaza, C. D'Souza, J. M. Barrutia, and C. Echebarria. 2017. Corporate environmental responsibility communication: Implications from CSR and green advertising research. In *Handbook of integrated CSR communication*, edited by S. Diehl, M. Karmasin, B. Mueller, R. Terlutter, and F. Weder, 377–92. New York: Springer International.
Herman, E., and R. McChesney. 1997. *The global media*. London: Cassell.
Hern, Alex. 2017. Google to overhaul advertising policies after growing boycott. *Guardian.com*, March 21. https://www.theguardian.com/technology/2017/mar/21/google-advertising-boycott-hateful-offensive-content
International Chamber of Commerce. 2011. *Framework for responsible environmental communications*. July. ICC Document Number 240-46/665.
IPCC. 2014a. Intergovernmental Panel on Climate Change: Fifth assessment report (AR5). http://www.ipcc.ch/activities/activities.shtml
IPCC. 2014b. *Mitigation of climate change*. Contribution of Working Group III to the Fifth Assessment Report of the Intergovernmental Panel on Climate

Change. Edited by O. Edenhofer, R. Pichs-Madruga, Y. Sokona, E. Farahani, S. Kadner, K. Seyboth, A. Adler, I. Baum, S. Brunner, P. and Eickemeier. Cambridge: Cambridge University Press. https://www.ipcc.ch/report/ar5/syr/

Jhally, S. 1997. *Advertising and the end of the world: The movie*. PDF transcript. Media Education Foundation. https://www.mediaed.org/assets/products/101/transcript_101.pdf

Jurado, D. G., and C. J. Maldonado. 2014. Advertising and climate change, relationships and divergences. *Interactions: Studies in Communication & Culture* 5 (1): 107–25.

Just, N., and M. Puppis. 2018. Moving beyond self-castigation: Let's reinvigorate communication policy research now! *Journal of Communication* 68 (2): 327–36.

Kahn, R. 2009. Producing crisis: Green consumerism as an ecopedagogical issue. In *Critical pedagogies of consumption: Living and learning in the shadow of the "shopocalypse"*, edited by J. Sandlin and P. McLaren, 47–57. New York: Routledge.

Kallis, G. 2011. In defense of degrowth. *Ecological Economics* 70 (5): 873–80.

Kallis, G. 2018. In defense of degrowth: Opinions and minifestos. Open Commons. https://www.academia.edu/32949817/In_defense_of_degrowth_-_Giorgos_Kallis.pdf

Kilbourne, W. E. 1995. Green advertising: Salvation or oxymoron? *Journal of Advertising* 24 (2): 7–20.

Kollewe, Julia. 2017. Google and Facebook bring in one-fifth of global ad revenue. *Guardian.com*, May 1. https://www.theguardian.com/media/2017/may/02/google-and-facebook-bring-in-one-fifth-of-global-ad-revenue

Laclau, E. 2014. Forward. In *International political economy of communication*, edited by C. Martens, E. Vivares, and R. W. McChesney, xi–xiii. New York: Palgrave Macmillan.

Latouche, S. 2010. Degrowth. *Journal of Cleaner Production* 18: 519–22.

Leiserowitz, A., E. Maibach, C. Roser-Renouf, G. Feinberg, and S. Rosenthal. 2016. Politics and global warming, Spring 2016. Yale University and George Mason University. New Haven: Yale Program on Climate Change Communication.

Leonard, Annie. 2011. *The story of stuff: How our obsession with stuff is trashing the planet, our communities, and our health—and a vision for change*. New York: Free Press.

Lewis, J., and T. Boyce. 2009. Climate change and the media: The scale of the challenge. In *Climate change and the media*, edited by J. Lewis and T. Boyce, 3–16. New York: Peter Lang.

Lieberstein, M., and B. Benjamin. 2010. Advertising and the environment: The new FTC Green Guides. *NSYBA Bright Ideas* 19 (3): 35–37.

Magdoff, F. 2014. An ecologically sound and socially just economy. *Monthly Review* 66 (4): 23–34.

Maxwell, R., and T. Miller. 2012a. *Greening the media*. New York: Oxford.

Maxwell, R., and T. Miller. 2012b. Greening starts with us. *New York Times*, September 23. http://www.nytimes.com/roomfordebate/2012/09/23/informations-environmental-cost/greening-starts-with-ourselves

McChesney, R. 2004. *The problem of the media*. New York: Monthly Review Press.

McChesney, R. 2013. *Digital disconnect*. New York: The New Press.

Morgan, T. 2015. Growing ourselves to death? Economic and ecological crises, the growth of waste, and the role of the media and cultural industries. *Human Geography* 8 (1): 68–81.

Mosco, V. 2014. *To the cloud: Big data in a turbulent world*. Boulder: Paradigm.

Mulholland, Rory. 2014. Grenoble to replace street advertising with trees and "community spaces." *The Telegraph*, November 24. https://www.telegraph.co.uk/news/worldnews/europe/france/11250670/Grenoble-to-replace-street-advertising-with-trees-and-community-spaces.html

Murdock, G. 2018. Media materialities: For a moral economy of machines. *Journal of Communication* 68 (2): 359–68.

Nuccitelli, Dana. 2017. Americans want a tax on carbon pollution but how to get one? *Guardian.com*, October 23. https://www.theguardian.com/environment/climate-consensus-97-per-cent/2017/oct/23/americans-want-a-tax-on-carbon-pollution-but-how-to-get-one

Nuccitelli, Dana. 2018. Canada passed a carbon tax that will give most Canadians more money. *Guardian.com*, October 26. https://www.theguardian.com/environment/climate-consensus-97-per-cent/2018/oct/26/canada-passed-a-carbon-tax-that-will-give-most-canadians-more-money

Oil Change International. 2016. The sky's limit: Why the Paris climate goals require a managed decline of fossil fuel production. Oil Change International, September. http://priceofoil.org/content/uploads/2016/09/OCI_the_skys_limit_2016_FINAL_2.pdf

OXFAM. 2015. Extreme carbon inequality: Why the Paris climate deal must put the poorest, lowest emitting and most vulnerable people first. Oxford Committee for Famine Relief (OXFAM). December 2. https://www.oxfam.org/sites/www.oxfam.org/files/file_attachments/mb-extreme-carbon-inequality-021215-en.pdf

Packard, V. 1960. *The waste makers*. New York: David McKay Company.

Rotzoll, K., J. Haefner, and C. Sandage. 1976. *Advertising in contemporary society: Perspectives toward understanding*. Columbus, OH: Grid.

Schutz, E. 2011. *Inequality and power: The economics of class*. New York: Routledge.

Shapiro, S. 2014. Poor people, poor planet: The psychology of how we harm and heal humanity and earth. In *Toward a socially responsible psychology for a global era*, edited by E. Mustakova-Possardt, M. Lyubansky, M. Basseches, and J. Oxenberg, 231–54. New York: Springer. https://doi.org/10.1007/978-1-4614-7391-6

Sheehan, Kim, and Deborah Morrison. 2018. Advertising leadership and climate change: The efficacy of industry professionals to address climate issues. *Advertising and Society Quarterly* 19 (1). https://muse-jhu-edu.ezproxy.fiu.edu/article/689165?

Shin, S., E. J. Ki, and W. G. Griffin. 2017. The effectiveness of fear appeals in "green" advertising: An analysis of creative, consumer, and source variables. *Journal of Marketing Communications* 23 (5): 1–20.

Springmann, M., D. Mason-D'Croz, S. Robinson, K. Wiebe, H. C. J. Godfray, and M. Rayner, et al. 2018. Health-motivated taxes on red and processed meat: A modelling study on optimal tax levels and associated health impacts. *PLoS ONE* 13 (11): e0204139. https://doi.org/10.1371/journal.pone.0204139

Storey, J. 2017. *Theories of consumption*. New York: Routledge.
Taylor, Astra. 2014. *The people's platform: Taking back power and culture in the digital age*. London: Fourth Estate.
United Nations Environment Programme. 2018. Emissions gap report. United Nations. http://wedocs.unep.org/bitstream/handle/20.500.11822/26896/EGR-KEYMESSAGES_2018.pdf?sequence=1&isAllowed=y
Vaughan, A. 2016. Google to be powered 100% by renewable energy from 2017. *Guardian.com*, December 6. https://www.theguardian.com/environment/2016/dec/06/google-powered-100-renewable-energy-2017
Waterson, J. 2018. UK advertising watchdog to crack down on sexist stereotypes. *Guardian.com*, December 13. https://www.theguardian.com/media/2018/dec/14/uk-advertising-watchdog-to-crack-down-on-sexist-stereotypers
Waterson, J. 2020. *Guardian* to ban advertising from fossil fuel firms. *Guardian.com*, January 29. https://www.theguardian.com/media/2020/jan/29/guardian-to-ban-advertising-from-fossil-fuel-firms-climate-crisis
Wolonick, R. 2012. Are Apple and Google in a race for North Carolina's "black gold"? *Minyanville*. http://www.minyanville.com/sectors/technology/articles/Are-Apple-and-Google-in-Race/12/7/2012/id/46453
World Federation of Advertisers. 2015. What is the WFA? http://www.wfanet.org/en/about-wfa

6. Digital Era Advertising, Surveillance, Exploitation, and Inequities

Acxiom.com. 2015. Axciom's data and technology have transformed marketing. http://www.Acxiom.com
Allevate. 2013. Face recognition: Profit, ethics, and privacy. *Allevate.com*. http://allevate.com/index.php/2013/01/07/face-recognition-in-retail-profit-ethics-and-privacy/
Andrejevic, M. 2011. Surveillance and alienation in the online economy. *Surveillance & Society* 8 (3): 278–87.
Andrejevic, M. 2015. Personal data: Blind spot of the "affective law of value"? *Information Society* 31 (1): 5–12.
Angwin, J. 2010. The web's new gold mine: Your secrets: A *Journal* investigation finds that one of the fastest-growing businesses on the Internet is the business of spying on consumers. *Wall Street Journal*, July 30. http://www.wsj.com/articles/SB10001424052748703940904575395073512989404
Armstrong, M., and J. Vickers. 2001. Competitive price discrimination. *RAND Journal of Economics* 32 (4): 1–27.
Arvidsson, A. 2011. Ethics and value in customer co-production. *Mass Marketing Theory* 11 (3): 261–78.
Arvidsson, A., and E. Colleoni. 2012. Value in informational capitalism and on the internet. *Information Society* 28 (3): 135–50.
Babe, R. E. 1994. The place of information in economics. In *Information and communication in economics*, edited by R. E. Babe, 41–68. Chicago: Springer Science and Business Media.
Bagdikian, B. 1989. *The new media monopoly*. Boston: Beacon Press.

Baran, P., and P. Sweezy. 2013. Theses on advertising. *Monthly Review* 65 (3): 34–42.
Benkler, Y. 2006. *The wealth of networks: How social production transforms markets and freedom.* New Haven: Yale University Press.
Borden, N. 1942. *The economic effects of advertising.* Chicago: Irwin.
Borgie, L. 2014. Trends in producer prices between e-commerce and brick-and-mortar retail trade establishments. Bureau of Labor Statistics 3, no. 18. http://www.bls.gov/opub/btn/volume-3/trends-in-producer-prices-between-e-commerce-and-brick-and-mortar-retail-trade-establishments.htm
Bridle, J. 2017. *The new dark age.* New York: Verso Books.
Browne, S. 2015. *Dark matters: On the surveillance of blackness.* Durham: Duke University Press.
Burston, J., N. Dyer-Witheford, and A. Hearn. 2010. Digital labour: Workers, authors, citizens. Special issue of *ephemera: theory and politics in organization* 10 (3/4): 214–21.
Caillaud, B., and R. De Nijs. 2014. Strategic loyalty reward in dynamic price discrimination. *Marketing Science* 33 (5): 725–42.
Carey, J. 1960. Advertising: An institutional approach. In *The role of advertising: A book of readings*, edited by C. Sandage and V. Fryburger, 3–17. Homewood, IL: Richard D. Irwin.
Census.gov. 2017. Quarterly retail e-commerce sales 2nd quarter, 2017. Census.gov. https://www.census.gov/retail/mrts/www/data/pdf/ec_current.pdf
Clark, L. 1998. Dating on the Net: Teens and the rise of "pure" relationships. In *Cybersociety 2.0: Revisiting computer-mediated communication and community*, edited by S. Jones, 159–83. Thousand Oaks, CA: Sage.
Danet, B. 1998. Text as mask: Gender, play, and performance on the Internet. In *Cybersociety 2.0: Revisiting computer-mediated communication and community*, edited by S. Jones, 129–58. Thousand Oaks, CA: Sage.
Dawson, M. 2005. *The consumer trap.* Urbana: University of Illinois Press.
Desjardins, F. 2017. Chart: Here's how 5 tech giants make their billions. *Virtual Capitalist.* http://www.visualcapitalist.com/chart-5-tech-giants-make-billions/
Dunn, J. 2017. The tech industry is dominated by 5 big companies—here's how each makes its money. *Business Insider*, May 26. http://www.businessinsider.com/how-google-apple-facebook-amazon-microsoft-make-money-chart-2017-5
Dyer, G. 1982. *Advertising as communication.* New York: Methuen and Co.
Dyer-Witheford, N. 1999. *Cyber-Marx: Cycles and circuits of struggle in high technology capitalism.* Chicago: University of Illinois Press.
Ellison, G., and S. Fisher-Ellison. 2009. Search, obfuscation, and price elasticities on the Internet. *Econometrica* 77 (2): 427–52.
Elmaghraby, W., and P. Keskinocak. 2003. Dynamic pricing in the presence of inventory considerations: Research overview, current practices, and future directions. *Management Science* 49 (10): 1287–1309.
eMarketer. 2014a. Total US ad spending to see largest increase since 2004: mobile advertising leads growth; will surpass radio, magazines and newspapers this year. *eMarketer*, July 2. http://www.emarketer.com/Article/Total-US-Ad-Spending-See-Largest-Increase-Since-2004/1010982
eMarketer. 2014b. Total US retail sales top $4.5 trillion in 2013, outpace GDP

growth. *eMarketer*, April 10. http://www.emarketer.com/Article/Total-US-Retail-Sales-Top-3645-Trillion-2013-Outpace-GDP-Growth/1010756
Eubanks, V. 2018. *Automating inequality: How high-tech tools profile, police, and punish the poor.* New York: St. Martin's Press.
Facefirst. 2015. Retail: Receive notifications on shoplifters, past employees, or felons. http://www.facefirst.com/services/retail
Faraone, R. 2011. Economy, ideology, and advertising. In *The handbook of political economy of communications*, edited by J. Wasco, G. Murdock, and H. Sousa, 187–205. Oxford: Blackwell.
Fisher, D. 1998. Rumoring theory and the Internet: A framework for analyzing the grassroots. *Social Science Computer Review* 16 (2): 158–68.
Foster, J. B., and R. W. McChesney. 2014. Surveillance capitalism: Monopoly-finance capital, the military-industrial complex, and the digital age. *Monthly Review* 66 (3): 1–31.
Fottrell, Q. 2012. Who would pay $5,000 to use Google? (you). http://blogs.smartmoney.com/advice/2012/01/25/who-would-pay-5000-to-use-google-you/
Fuchs, C. 2010. Labor in informational capitalism and on the Internet. *Information Society* 26 (3): 179–96.
Fuchs, C. 2012a. Political economy of privacy on Facebook. *Television & New Media* 13 (2): 139–59.
Fuchs, C. 2012b. With or without Marx? With or without capitalism? A Rejoinder to Adam Arvidsson and Eleanor Colleoni. *tripleC: Communication, Capitalism & Critique* 10 (2): 633–45.
Fuchs, C. 2015. *Culture and economy in the age of social media.* New York: Routledge.
Galbraith, J. 1969. *The affluent society.* Boston: Houghton Mifflin.
Garcia, Ahiza. 2015. Mark Zuckerberg: Internet access can eradicate extreme poverty. *Cnn.com*, September 27. http://money.cnn.com/2015/09/26/news/mark-zuckerberg-united-nations-poverty-internet/index.html
Ghose, A., and K. W. Huang. 2009. Personalized pricing and quality customization. *Journal of Economics & Management Strategy* 18 (4): 1095–1135.
Goldstein, P. 1994. *Copyright's highway: From Gutenberg to the celestial jukebox.* New York: Hill and Wang.
Gomes, R., and A. Pavan. 2017. Price customization and targeting in many-to-many matching markets. Northwestern University. http://faculty.wcas.northwestern.edu/~apa522/Targeting-oct-19.pdf
Hannak, A., G. Soeller, D. Lazer, A. Mislove, and C. Wilson. 2014. Measuring price discrimination and steering on e-commerce web sites. In *Proceedings of the 2014 conference on internet measurement conference*, 305–318.
Hesmondhalgh, D. 2010. User-generated content, free labour and the cultural industries. *ephemera* 10 (3/4): 267–84.
Hill, K. 2015. You're being secretly traced with facial recognition, even in church. *Fusion*, June 23. http://fusion.net/story/154199/facial-recognition-no-rules/
Hirschman, D. 2011. On the road to riches: Those under 35 with $100K household income. *Advertising Age*, May 22. http://adage.com/article/adagestat/ll-rich-35-100k-household-income/227671/
Holm, N. 2017. Art for fun and profit: The political aesthetics of advertising. In

Explorations in critical studies of advertising, edited by J. Hamilton, R. Bodle, and E. Korin, 97–109. New York: Routledge.

Homa, K. 2015. Gotcha: Your willingness-to-pay is showing. *The Homa Files* blog. https://kenhoma.wordpress.com/2014/12/18/gotcha-your-willingness-to-pay-is-showing-3/#more-27632

Hubbard, K. 2010. Digits: How RapLeaf mines data online. *Wall Street Journal*, October 25. http://www.wsj.com/video/digits-how-rapleaf-mines-data-online/6B7F29FE-4A2C-4619-BCB7-CCCE5EB35F62.html

International Air Transport Association. 2015. News brief: IATA announces new NDC pilots and deployments. https://www.iata.org/pressroom/pr/Pages/2015-06-08-06.aspx

Investopedia. 2017. What are some current examples of oligopolies? May 8. https://www.investopedia.com/ask/answers/121514/what-are-some-current-examples-oligopolies.asp

IPG Mediabrands. 2014. MAGNA GLOBAL forecasts global advertising revenues to grow by +4.8% to $536 billion in 2015. http://news.ipgmediabrands.com/article_display.cfm?article_id=1666

Klock, M. 2002. Unconscionability and price discrimination. *Tennessee Law Review* 69: 317–81.

Knight Frank Research. 2016. The wealth report. http://content.knightfrank.com/research/83/documents/en/wealth-report-2016-3579.pdf

Kolko, B., and E. Reid. 1998. Dissolution and fragmentation: Problems in on-line communities. In *Cybersociety 2.0: Revisiting computer-mediated communication and community*, edited by S. Jones, 129–58. Thousand Oaks, CA: Sage.

LBBW. 2017. The right place at all times. https://www.lbbw.de/enhub/artikelseite/artikel_detailseite_55498.html

Lewis, J. 2013. *Beyond consumer capitalism: Media and the limits of imagination*. Boston: Polity Press.

Lohr, S. 2009. The data that turns browsing to buying. *New York Times*, December 5. http://www.nytimes.com/2009/12/06/business/06unboxed.html?_r=0

Manyika, J., M. Chui, B. Brown, J. Bughin, R. Dobbs, C. Roxburgh, and A. Hung. 2011. Big data: The next frontier for innovation, competition, and productivity. http://www.mckinsey.com/insights/business_technology/big_data_the_next_frontier_for_innovation

Marx, K. 1973. *Grundrisse der Kritik der Politischen Oekonomie* (Outlines of the critique of political economy). Translated by Martin Nicolaus. Harmondsworth: Penguin.

McChesney, R. W. 2004. *The problem of the media*. New York: Monthly Review Press.

McChesney, R. W. 2013. *Digital disconnect: How capitalism is turning the internet against democracy*. New York: New Press.

McKinsey and Company. 2014. Global media report: Global industry overview. http://www.mckinsey.com/industries/media-and-entertainment/our-insights/global-media-report-2014

Melody, W. 2006. Markets and policies in new knowledge economies. *LIRNE.NET*. https://idl-bnc-idrc.dspacedirect.org/bitstream/handle/10625/41875/129520.pdf?sequence=1

Morishima, M. 1973. *Marx's economics: A dual theory of value and growth.* Cambridge: Cambridge University Press.
Morley, K. 2017. Exclusive: End of fixed prices within five years as supermarkets adopt electronic price tags. *Telegraph*, June 24. http://www.telegraph.co.uk/news/2017/06/24/exclusive-end-fixed-prices-within-five-years-supermarkets-adopt/
Mosco, V. 2014. *To the cloud: Big data in a turbulent world.* Boulder, CO: Paradigm.
Mosco, V. 2017. *Becoming digital: Toward a post-internet society.* United Kingdom: Emerald Publishing.
Negroponte, N. 1996. *Being digital.* New York: Vintage Books.
Newman, N. 2014. The costs of lost privacy: Consumer harm and rising economic inequality in the age of Google. *William Mitchell Law Review* 40 (2): 850–88.
Newman, N. 2015. Data justice: Taking on big data as an economic justice issue. http://www.datajustice.org/sites/default/files/Data%20Justice-%20Taking%20on%20Big%20Data%20as%20an%20Economic%20Justice%20Issue.pdf
Nichols, J., and R. W. McChesney. 2013. *Dollarocracy: How the money and media election complex is destroying America.* New York: Nation Books.
Noble, Safiya Umoja. 2018. *Algorithms of oppression: How search engines reinforce racism.* New York: New York University Press.
Norris, V. 1989. Advertising history—according to the textbooks. In *Advertising in society: Classic and contemporary readings on advertising's role in society*, edited by Roxanne Hovland and Gary Wilcox, 99–118. Lincolnwood, IL: NTC Business Books.
Obermiller, C., D. Arnesen, and M. Cohen. 2012. Customized pricing: Win-win or end run? *Drake Management Review* 1 (2): 12–28.
O'Connor, P. 2014. Political ads take targeting to the next level. *Wall Street Journal*, November 20. http://www.wsj.com/articles/political-ads-take-targeting-to-the-next-level-1405381606
Office of Fair Trading. 2010. Online targeting of advertising and prices: A market study. http://webarchive.nationalarchives.gov.uk/20140402142426/http:/www.oft.gov.uk/shared_oft/business_leaflets/659703/OFT1231.pdf
O'Neil, C. 2017. *Weapons of math destruction: How big data increases inequality and threatens democracy.* New York: Broadway Books.
Pariser, E. 2012. *The filter bubble: How the new personalized web is changing what we read and how we think (reprint edition).* New York: Penguin Books.
Park, D. J. 2009. *Conglomerate rock: The music industry's quest to divide music and conquer wallets.* Lanham, MD: Lexington Books.
Park, D. J. 2017. Toward an ecological critique of advertising. In *Explorations in critical studies of advertising*, edited by James Hamilton, R. Bodle, and E. Korin, 67–80. New York: Routledge.
Pavlik, J. 1998. *New media technology: Cultural and commercial perspectives.* 2nd ed. Boston: Allyn and Bacon.
Perlberg, S. 2014. Targeted ads? TV can do that now too. *Wall Street Journal*, November 20. http://www.wsj.com/articles/targeted-ads-tv-can-do-that-now-too-1416506504
Personalizing online prices: How deep are your pockets? Businesses are offered

software that spots which customers will pay more. 2012. *Economist.com*, June 30. http://www.economist.com/node/21557798

Reich, R. 2015. Bernie Sanders tells the truth: Former Clinton labor secretary Robert Reich on his surging campaign. *Democracynow!*, September 30. http://www.democracynow.org/2015/9/30/bernie_sanders_tells_the_truth_former?autostart=true&get_clicky_key=suggested_next_story

Reinartz, W. 2002. Customizing prices in online markets. *Emerging Issues in Management* 1: 55–65.

Rezabakhsh, B., D. Bornemann, U. Hansen, and U. Schrader. 2006. Consumer power: A comparison of the old economy and the internet. *Journal of Consumer Policy* 29: 3–36.

Ritzer, G., and Jurgenson, N. 2010. Production, consumption, prosumption: The nature of capitalism in the age of the digital "prosumer." *Journal of Consumer Culture* 10 (1): 13–36.

Rocca, M. 2015. Henry Ford's innovation nation with Mo Rocca: Facial recognition and shopping. http://cbsdreamteam.com/the-henry-fords-innovation-nation/episodes/facial-recognition-and-shopping/

Rosenblat, A., R. Randhava, D. Boyd, S. Gangadharan, and C. Yu. 2014. Data and civil rights: Consumer finance primer. Data & Society Research Institute. http://www.datacivilrights.org/pubs/2014-1030/Finance.pdf

Rotzoll, K., J. Haefner, and C. Sandage. 1976. *Advertising in contemporary society: Perspectives toward understanding*. Columbus: Grid.

Rubens, P. 2014. Facial recognition: Shop where everyone knows your name. *BBC.com*, December 8. http://www.bbc.com/news/business-30219820

Saad-Filho, A. 2003. Introduction. In *Anti-Capitalism: A Marxist introduction*, edited by A. Saad-Filho, 1–16. London: Pluto Press.

Sahay, A. 2007. How to reap higher prices with dynamic pricing. *MIT Sloan Management Review*, Summer. http://sloanreview.mit.edu/article/how-to-reap-higher-profits-with-dynamic-pricing/

Salinas, B. 2013. High-end stores use facial recognition tools to spot VIPs. *NPR.org*, July 21. http://www.npr.org/blogs/alltechconsidered/2013/07/21/203273764/high-end-stores-use-facial-recognition-tools-to-spot-vips

Schiller, D. 1996. *Theorizing communication: A history*. New York: Oxford University Press.

Schiller, D. 2014. *Digital depression: Information technology and economic crisis*. Urbana: University of Illinois Press.

Schutz, E. 2011. *Inequality and power: The economics of class*. New York: Routledge.

Shapiro, A. 1999. *The control revolution: How the internet is putting individuals in charge and changing the world we know*. New York: PublicAffairs.

Shapiro, C., and H. Varian. 1999. *Information rules: A strategic guide to the network economy*. Cambridge, MA: Harvard Business School Press.

Shiller, B. 2013. First degree price discrimination using big data. Working paper series. Brandeis University. http://www.brandeis.edu/departments/economics/RePEc/brd/doc/Brandeis_WP58.pdf

Singer, N. 2012. Mapping, and sharing, the consumer genome. *New York Times*, June 16. http://www.nytimes.com/2012/06/17/technology/acxiom-the-quiet-giant-of-consumer-database-marketing.html

Splichal, S. 2009. "New" media, "old" theories: Does the (national) public melt into the air of global governance? *European Journal of Communication* 24 (4): 391–405.

Srinivasan, J., M. Finn, and M. Ames. 2017. Information determinism: The consequences of the faith in information. *Information Society* 33 (1): 13–22.

Srnicek, Nick. 2017. *Platform capitalism*. Cambridge: Polity Press.

Stiglitz, J. 2002. Information and the change in the paradigm in economics. *American Economic Review* 92 (3): 460–501.

Statista. 2017. Share of global retail e-commerce sales. *Statista.com*. https://www.statista.com/statistics/243699/share-of-global-retail-e-commerce-sales-usa/

Storey, J. 2017. *Theories of consumption*. New York: Routledge.

Sunstein, C. 2001. *Republic.com*. Princeton: Princeton University Press.

Tanner, A. 2014. Different customers, different prices, thanks to big data. *Forbes.com*, March 26. http://www.forbes.com/forbes/welcome/?toURL=http://www.forbes.com/sites/adamtanner/2014/03/26/different-customers-different-prices-thanks-to-big-data/&refURL=https://search.yahoo.com/&referrer=https://search.yahoo.com/

Taplin, Jonathan. 2017. *Move fast and break things: How Facebook, Google, and Amazon cornered culture and undermined democracy*. New York: Little, Brown.

Taylor, Astra. 2014. *The people's platform: Taking back power and culture in the digital age*. London: Fourth Estate.

Thielman, S. 2015. World's biggest tech companies get failing grade on data-privacy rights. *Guardian.com*, November 3. https://www.theguardian.com/technology/2015/nov/03/data-protection-failure-google-facebook-ranking-digital-rights

Tuchman, A., H. Nair, and P. Gardete. 2015. Complementarities in consumption and the consumer demand of advertising. Stanford University. http://web.stanford.edu/~tuchman/TuchmanNairGardete_ComplementaritiesAdvertisingConsumption.pdf

Turow, J. 1997. *Breaking up America: Advertisers and the new media world*. Chicago: University of Chicago Press.

Turow, J. 2011. *The daily you: How the new advertising industry is defining your identity and your worth*. New Haven: Yale University Press.

Turow, J., L. Feldman, and K. Meltzer. 2005. Open to exploitation: America's shoppers online and offline. Annenberg Public Policy Center of the University of Pennsylvania. http://repository.upenn.edu/asc_papers/35

U.S. Government, Subcommittee on Networking and Information Technology Research and Development. 2016. *Federal big data research and development strategic plan*. Washington, DC. https://www.nitrd.gov/PUBS/bigdatardstrategicplan.pdf

U.S. marketing spending exceeded $1 trillion in 2005. 2006. Metrics Business and Marketing Intelligence, June 26. http://metrics2.com.

Vaidhyanathan, Siva. 2018. *Antisocial media: How Facebook disconnects us and undermines democracy*. London: Oxford University Press.

Valentino-Devries, J., J. Singer-Vine, and A. Soltani. 2012. Websites vary prices, deals based on users. *Wall Street Journal*, December 23. http://on.wsj.com/Tj1W2V

Web-based targeted advertising in a brick-and-mortar retail establishment using online customer information. 2008. Microsoft Corporation. US 20080004951 A1. https://www.google.com/patents/US20080004951?dq=facial+recognition+advertising&hl=en&sa=X&ei=WcJoVamnOsmTsQTDtYOQDQ&ved=0CFwQ6AEwCQ

Winseck, D. 2016. Reconstructing the political economy of communication for the digital media age. *Political Economy of Communication* 4 (2): 73–114.

World Bank. 2015. Final consumption expenditure, etc. (current US$). World Bank. https://data.worldbank.org/indicator/NE.CON.TETC.CD

World Federation of Advertisers. 2015. What is the WFA? http://www.wfanet.org/en/about-wfa

Wright, E. O. 1997. *Class counts: Comparative studies in class analytics*. Cambridge: Cambridge University Press.

Wyatt, E. 2013. Most of U.S. is wired, but millions aren't plugged in. *New York Times*, August 18. http://www.nytimes.com/2013/08/19/technology/a-push-to-connect-millions-who-live-offline-to-the-internet.html?_r=0

Zuboff, S. 2019. *The age of surveillance capitalism: The fight for a human future at the new frontier of power*. New York: PublicAffairs.

Zwick, D., S. K. Bonsu, and A. Darmody. 2008. Putting consumers to work: Co-creation and new marketing govern-mentality. *Journal of Consumer Culture* 8 (2): 163–96.

7. Challenges to Sustainability in the Digital Era

Asara, V., I. Otero, F. Demaria, and E. Corbera. 2015. Socially sustainable degrowth as a social-ecological transformation: Repoliticizing sustainability. *Sustainability Science* 10 (3): 375–84.

Baldé, C. P., V. Forti, V. Gray, R. Kuehr, and P. Stegmann. 2017. The global e-waste monitor—2017. United Nations University, International Telecommunication Union, and International Solid Waste Association. Bonn/Geneva/Vienna. https://collections.unu.edu/eserv/UNU:6341/Global-E-waste_Monitor_2017__electronic_single_pages_.pdf

Buchheit, P. 2015. New 2015 wealth data: U.S. inequality at its ugliest: According to new Credit Suisse data, nearly 50 million of America's 243 million adults are part of the world's poorest 10%. *Commondreams.org*, October 19. http://www.commondreams.org/views/2015/10/19/new-2015-wealth-data-us-inequality-its-ugliest

Carroll, C. D., and L. H. Summer. 1991. Consumption growth parallels income growth: Some new evidence. In *National savings and economic performance*, edited by B. D. Berhheim and J. B. Shoven, 305–47. Chicago: University of Chicago Press.

Chancel, L., and T. Piketty. 2015. Carbon and inequality: From Kyoto to Paris. Trends in the global inequality of carbon emissions (1998–2013) and prospects for an equitable adaptation fund. Paris School of Economics. http://piketty.pse.ens.fr/files/ChancelPiketty2015SummaryEN.pdf

Christensen, Miyase, and Annika E. Nilsson. 2018. Media, communication, and the environment in precarious times. *Journal of Communication* 68 (2) 1: 267–77.

Collins, C., and J. Hoxie. 2015. Billionaire bonanza report: The Forbes 400 and the rest of us. Institute for Policy Studies. http://www.ips-dc.org/billionaire-bonanza/

Confessore, N., S. Cohen, and K. Yourish. 2015. Buying power. *New York Times*, October 10. http://www.nytimes.com/interactive/2015/10/11/us/politics/2016-presidential-election-super-pac-donors.html

Congressional Budget Office. 2007. Trade-offs in allocating allowances for CO2 emissions. Economic and Budget Issue Brief. 1–8. http://cbo.gov

Data Center Journal. 2013. Industry perspective: Energy efficiency and renewable sources for the data center. http://www.datacenterjournal.com/facilities/industry-perspective-energy-efficiency-and-renewable-sources-for-the-data-center/

Economist. 2020. Democracy index 2020. http://economist.com/graphic-detail/2020/01/22/global-democracy-has-another-bad-year

Galbraith, James K. 2012. *Inequality and instability: A study of the world economy before the great crisis*. New York: Oxford University Press.

Garces, M., and S. Rendall. 2012. Media not concerned about the very poor: Fairness and Accuracy in Media. *Extra!*, September 7: 5–6.

Gartner. 2007. Gartner estimates ICT industry accounts for 2 percent of global CO2 emissions. *Gartner.com*, April 26. http://www.gartner.com/newsroom/id/503867

Glanz, James. 2012. Power, pollution, and the internet. *New York Times*, September 22. http://www.nytimes.com/2012/09/23/technology/data-centers-waste-vast-amounts-of-energy-belying-industry-image.html

Global video streaming market expected to grow with a CAGR of 18.3% over the forecast period, 2019–2026—ResearchAndMarkets.com. 2020. *Businesswire.com*, February 25. https://www.businesswire.com/news/home/20200225005696/en/Global-Video-Streaming-Market-Expected-Grow-CAGR

Goodwin, N., J. Nelson, A. Ackerman, and T. Weisskopf. N.d. Consumption and the consumer society. Global Development and Environment Institute, Tufts University. http://www.ase.tufts.edu/gdae/education_materials/modules/consumption_and_the_consumer_society.pdf

Greenpeace. 2019. Clicking clean Virginia: The dirty energy powering data center alley. https://www.greenpeace.org/usa/reports/click-clean-virginia/

Greenpeace International. 2010. Making IT green: Cloud computing and its contributions to climate change. http://www.greenpeace.org/international/Global/international/planet-2/report/2010/3/make-it-green-cloud-computing.pdf

Grossman, Elizabeth. 2010. Tackling high-tech trash: The e-waste explosion and what we can do about it. Demos, November 22. http://www.demos.org/publication/tackling-high-tech-trash-e-waste-explosion-what-we-can-do

Hackett, R. A. 2016. Dallas Smythe Award keynote lecture. *Democratic Communiqué* 27: 1–14.

InfluenceMap. 2019. Big oil's real agenda on climate change. March. https://influencemap.org/report/How-Big-Oil-Continues-to-Oppose-the-Paris-Agreement-38212275958aa21196dae3b76220bddc

Information Observatory. 2016. Billionaires who owe their wealth to information. *Information Observatory*, May 1. http://informationobservatory.info/2016/05/01/billionaires-who-owe-their-wealth-to-information/

IPCC. 2014a. Intergovernmental Panel on Climate Change: Fifth assessment report (AR5). http://www.ipcc.ch/activities/activities.shtml

IPCC. 2014b. *Mitigation of climate change.* Contribution of Working Group III to the Fifth Assessment Report of the Intergovernmental Panel on Climate Change. Edited by O. Edenhofer, R. Pichs-Madruga, Y. Sokona, E. Farahani, S. Kadner, K. Seyboth, A. Adler, I. Baum, S. Brunner, P. Eickemeier, et al. Cambridge: Cambridge University Press. https://www.ipcc.ch/report/ar5/syr/

Jokinen, P., P. Malaska, and J. Kaivo-oja. 1998. The environment in an "information society": A transition stage towards a more sustainable development? *Futures* 30 (6): 485–98.

Knight Frank Research. 2016. The wealth report. http://content.knightfrank.com/research/83/documents/en/wealth-report-2016-3579.pdf

Kollewe, Julia. 2016. World's super rich keep buying up luxury goods in face of wealth decline. *Guardian.com*, March 1. http://www.theguardian.com/business/2016/mar/02/global-super-rich-luxury-goods-yachts-wealth-report-knight-frank

Kuehr, R., and E. Williams. 2003. *Computers and the environment: Understanding and managing their impacts.* Dordrecht: Kluwer Academic Publishers.

Kutsch, T. 2015. 500 US firms keep $2.1 trillion in tax havens, study finds. *Al-Jazeera.com*, October 6. http://america.aljazeera.com/articles/2015/10/6/top-us-companies-keep-21-trillion-in-tax-havens-abroad.html

Lebow, Victor. 1955. Price competition in 1955. *Journal of Retailing* 16 (1): 6–10.

Leonard, Annie. 2011. *The story of stuff: How our obsession with stuff is trashing the planet, our communities, and our health—and a vision for change.* New York: Free Press.

Maxwell, R., and T. Miller. 2012a. Greening starts with us. *New York Times*, September 24. http://www.nytimes.com/roomfordebate/2012/09/23/informations-environmental-cost/greening-starts-with-ourselves

Maxwell, R., and T. Miller. 2012b. *Greening the media.* New York: Oxford.

McChesney, R. W. 2013. *Digital disconnect: How capitalism is turning the internet against democracy.* New York: New Press.

McKernan, S. M., C. Ratcliffe, and S. Cellini. 2009. Transitioning in and out of poverty. Urban Institute, September 10. http://www.urban.org/sites/default/files/alfresco/publication-pdfs/411956-Transitioning-In-and-Out-of-Poverty.PDF

Milne, S. 2015. The Davos oligarchs are right to fear the world they've made. *Guardian.com*, January 22. https://www.theguardian.com/commentisfree/2015/jan/22/davos-oligarchs-fear-inequality-global-elite-resist

Mosco, V. 2014. *To the cloud: Big data in a turbulent world.* Boulder, CO: Paradigm Publishers.

Neate, Rupert. 2020. The super-rich: Another 31,000 people join the ultra-wealthy. *Guardian.com*, March 4. https://www.theguardian.com/news/2020/mar/04/the-super-rich-another-31000-people-join-the-ultra-wealthy-elite

Newman, N. 2014. The costs of lost privacy: Consumer harm and rising economic inequality in the age of Google. *William Mitchell Law Review* 40 (2): 850–88.

Nichols, J., and R. W. McChesney. 2013. *Dollarocracy: How the money and media election complex is destroying America.* New York: Nation Books.

OXFAM. 2015. World's richest 10% produce half of carbon emissions while poorest 3.5 billion account for just a tenth. Oxfam.org. https://www.oxfam.org/en/pressroom/pressreleases/2015-12-02/worlds-richest-10-produce-half-carbon-emissions-while-poorest-35

Packard, V. 1960. *The waste makers*. New York: David McKay Company.

Park, D. J. 2009. *Conglomerate rock: The music industry's quest to divide music and conquer wallets*. Lanham, MD: Lexington Books.

Piketty, T. 2014. *Capital in the twenty-first century*. Cambridge, MA: Belknap Press of Harvard University Press.

Rauch, Jennifer. 2018. *Slow media: Why "slow" is satisfying, sustainable, and smart*. New York: Oxford University Press.

Rushe, Dominic. 2018. US cities and states give big tech $9.3bn in subsidies in five years. *Guardian.com*, July 2. https://www.theguardian.com/cities/2018/jul/02/us-cities-and-states-give-big-tech-93bn-in-subsidies-in-five-years-tax-breaks

Saez, E., and G. Zucman. 2014. Wealth inequity in the United States since 1913: Evidence from capitalized income tax data. National Bureau of Economic Research. Working paper 20625. http://eml.berkeley.edu/~saez/saez-zucmanNBER14wealth.pdf

Schiller, D. 2014. *Digital depression: Information technology and economic crisis*. Urbana: University of Illinois Press.

Schutz, E. 2011. *Inequality and power: The economics of class*. New York: Routledge.

Sterne, J. 2007. Out with the trash: On the future of new media. In *Residual media*, edited by C. R. Acland, 16–31. Minneapolis: University of Minnesota Press.

Stiglitz, J. 2002. Information and the change in the paradigm in economics. *American Economic Review* 92 (3): 460–501.

Taylor, Astra. 2014. *The people's platform: Taking back power and culture in the digital age*. London: Fourth Estate.

Tuttle, Brad. 2013 The splurge surge: Luxury spending on the rise. *Time Magazine*, May 1. http://business.time.com/2013/05/01/the-splurge-surge-luxury-spending-on-the-rise/

US Energy Information Administration. 2016. EIA projects 48% increase in world energy consumption by 2040. www.eia.gov/todayinenergy/detail.cfm?id=26212

White House. 2015. Big data and differential pricing. Executive Office of the President of the United States. https://www.whitehouse.gov/sites/default/files/docs/Big_Data_Report_Nonembargo_v2.pdf

Wicker, Stephan, and Kolbeinn Karlsson. 2017. Internet advertising: Technology, ethics, and a serious difference of opinion. *Communications of the ACM* 60 (10): 70–77.

Williams, E. 2004. Energy intensity of computer manufacturing: Hybrid assessment combining process and economic input-output methods. *Environmental Science and Technology* 38 (22): 6166–74.

Wingfield, N. 2016. Microsoft earnings are up, cushioned by its cloud business. *New York Times*, July 19. http://www.nytimes.com/glogin?URI=http%3A%2F%2Fwww.nytimes.com%2F2016%2F07%2F20%2Ftechnology%2Fmicrosoft-earnings-are-up-cushioned-by-its-cloud-business.html%3F_r%3D0

Wolff, E. 2012. The asset price meltdown and the wealth of the middle class. NBER working paper no. 18559. http://ecineq.org

Zuboff, S. 2019. *The age of surveillance capitalism: The fight for a human future at the new frontier of power.* New York: PublicAffairs.

8. Challenges to US Climate Change Journalism

Anderson, A. 2009. Media, politics and climate change: Towards a new research agenda. *Sociology Compass* 3 (2): 166–182.
Anderson, A. 2011. Sources, media, and modes of climate change communication: the role of celebrities. *Wiley Interdisciplinary Reviews: Climate Change* 2 (4): 535–46.
Arnold, Amanda. 2019. As many as 2,100 people have lost media jobs in the past two weeks. *The Cut*, February 1. https://www.thecut.com/2019/02/vice-layoffs-buzzfeed-huffpost-media.html
Bennett W. L. 2002. *News: The politics of illusion.* New York: Longman.
Boykoff, M. T. 2007. From convergence to contention: United States mass media representations of anthropogenic climate change science. *Transactions of the Institute of British Geographers* 32 (4): 477–89.
Boykoff, M. T. 2008. Lost in translation? United States television news coverage of anthropogenic climate change, 1995–2004. *Climatic Change* 86 (1–2): 1–11.
Boykoff, M. T. 2013. Public enemy no. 1? Understanding media representations of outlier views on climate change. *American Behavioral Scientist* 57 (6): 796–817.
Boykoff, M. T., and J. M. Boykoff. 2007. Climate change and journalistic norms: A case-study of US mass-media coverage. *Geoforum* 38 (6): 1190–1204.
Boykoff, M. T., and J. Farrell. 2020. Climate change countermovement organizations and media attention in the United States. In *Climate change denial and public relations: Strategic communication and interest groups in climate inaction*, edited by Núria Almiron and Jordi Xifra, 121–39. New York: Routledge.
Boykoff, M. T., and S. R. Rajan. 2007. Signals and noise. *EMBO Reports* 8 (3): 207–11.
Boykoff, M. T., and J. T. Roberts. 2007. Media coverage of climate change: Current trends, strengths, weaknesses. *Human Development Report 2007/2008*, 1–53. New York: United Nations Development Programme.
Burgess, J. 1990. The production and consumption of environmental meanings in the mass media: A research agenda for the 1990s. *Transactions of the Institute of British Geographers*, 139–61.
Brüggemann, M., and S. Engesser. 2014. Between consensus and denial climate journalists as interpretive community. *Science Communication* 36 (4): 399–427.
Carmody, K. 1995. It's a jungle out there: Environmental journalism in the age of backlash. *Columbia Journalism Review* 34 (1): 40–45.
Carvalho, A. 2010. Media (ted) discourses and climate change: A focus on political subjectivity and (dis) engagement. *Wiley Interdisciplinary Reviews: Climate Change* 1 (2): 172–79.
Carvalho, A., and J. Burgess. 2005. Cultural circuits of climate change in UK broadsheet newspapers, 1985–2003. *Risk Analysis* 25 (6): 1457–69.
Cooper, Evlondo. 2019. Climate change is fueling storms like Hurricane Dorian: But you wouldn't know that from watching broadcast TV news. Mediamatters.

org. https://www.mediamatters.org/broadcast-networks/how-broadcast-news-failed-climate-change-and-hurricane-dorian

Democracy Now! 2016. As Hurricane Matthew lashes U.S., are TV networks the last bastion of climate denial? October 7. http://www.democracynow.org/shows/2016/10/7

Deuze, M., P. Elefante, and B. Steward. 2010. Media work and the recession. *Popular Communication* 8 (3): 226–31.

Dunwoody, S., and H. P. Peters. 1992. Mass media coverage of technological and environmental risks: A survey of research in the United States and Germany. *Public Understanding of Science* 1 (2): 199–230.

Entman, R. M. 1993. Framing: Toward clarification of a fractured paradigm. *Journal of Communication* 43 (4): 51–58.

Ereaut, G., and N. Segnit. 2006. Warm words: How we are telling the climate story and can we tell it better. Institute for Public Policy Research. https://www.ippr.org/publications/warm-wordshow-are-we-telling-the-climate-story-and-can-we-tell-it-better

Fahy, D., and M. C. Nisbet. 2011. The science journalist online: Shifting roles and emerging practices. *Journalism* 12 (7): 778–93.

Feldman, L., E. W. Maibach, C. Roser-Renouf, and A. Leiserowitz. 2012. Climate on cable: The nature and impact of global warming coverage on Fox News, CNN, and MSNBC. *International Journal of Press/Politics* 17 (1): 3–31.

Feldman, L., T. A. Myers, J. D. Hmielowski, and A. Leiserowitz. 2014. The mutual reinforcement of media selectivity and effects: Testing the reinforcing spirals framework in the context of global warming. *Journal of Communication* 64 (4): 590–611.

Fuchs, C. 2014. *Digital labor and Karl Marx*. London: Routledge.

Gabbatt, Adam. 2020. US newspapers face "extinction-level" crisis as Covid-19 hits hard. *Guardian.com*, April. https://www.theguardian.com/media/2020/apr/09/coronavirus-us-newspapers-impact

Gavin, N. T., L. Leonard-Milsom, and J. Montgomery. 2011. Climate change, flooding and the media in Britain. *Public Understanding of Science* 20 (3): 422–38.

Gelbspan, R. 2005. Snowed: Though global climate change is breaking out all around us, the U.S. news media has remained silent. *Mother Jones*, May. http://www.motherjones.com/politics/2005/05/snowed

Gibson, T. A., R. T. Craig, A. C. Harper, and J. M. Alpert. 2015. Covering global warming in dubious times: Environmental reporters in the new media ecosystem. *Journalism*, January 4. https://doi.org/10.1177/1464884914564845

Good, J. E. 2008. The framing of climate change in Canadian, American, and international newspapers: A media propaganda model analysis. *Canadian Journal of Communication* 33 (2): 233.

Grant, M. 2016. Media unwilling to investigate cause of record-breaking winter heat. *EXTRA!* 29 (2).

Hackett, R. A. 2016. Dallas Smythe Award keynote lecture. *Democratic Communiqué* 27: 1–14.

Happer, C., and G. Philo. 2016. New approaches to understanding the role of the news media in the formation of public attitudes and behaviours on climate change. *European Journal of Communication* 31 (2): 136–51.

Harbinson, R., R. Mugara, and A. Chawla. 2006. Whatever the weather: Media attitudes to reporting climate change. Panos Institute Report London. http://cdn.agilitycms.com/wacc-global/resources/climate-justice/6.%20PANOS-Whatever-the-weather(2006).pdf

Henderson-Sellers, A. 1998. Climate whispers: Media communication about climate change. *Climatic Change* 40 (3–4): 421–56.

Hiles, S. S., and A. Hinnant. 2014. Climate change in the newsroom journalists' evolving standards of objectivity when covering global warming. *Science Communication*, May 19. https://doi.org/10.1177/1075547014534077

Jackson, Janine. 2019. GateHouse's takeover of Gannett: Bad news for journalism and the planet. Fair.org, August 23. https://fair.org/home/gatehouses-takeover-of-gannett-bad-news-for-journalism-and-the-planet/?awt_l=Ivb6m&awt_m=iAvnl6VOc0OI_TQ

Johnson, Adam. 2017. In a dozen interviews, media never bothered asking President Trump about climate change. Fair.org, June 3. https://fair.org/home/in-a-dozen-interviews-media-never-bothered-asking-president-trump-about-climate-change/

Kalhoefer, Kevin. 2017. How broadcast networks covered climate change in 2016. *Mediamatters.org*, March 23. https://www.mediamatters.org/research/2017/03/23/how-broadcast-networks-covered-climate-change-2016/215718

Lozada, Hector, Gary Kritz, and Alma Mintu-Wimsatt. 2017. Let's (not) talk about climate change: American consumers and global warming. Marketing Management Association Proceedings.

MacDonald, Ted. 2018. Major broadcast TV networks mentioned climate change just once during two weeks of heat-wave coverage. *Mediamatters.org*, July 12. https://www.mediamatters.org/blog/2018/07/12/Major-broadcast-TV-networks-mentioned-climate-change-just-once-during-two-weeks-of-heat-wa/220651

Macdonald, Ted, and Lisa Hymas. 2019. How broadcast TV networks covered climate change in 2018. *Mediamatters.org*, March 11. https://www.mediamatters.org/research/2019/03/11/How-broadcast-TV-networks-covered-climate-change-in-2018/223076

Mazur, A., and J. Lee. 1993. Sounding the global alarm: Environmental issues in the US national news. *Social Studies of Science* 23 (4): 681–720.

McChesney, R. W. 2013. *Digital disconnect: How capitalism is turning the internet against democracy*. New York: New Press.

McChesney, R. W. 2019. Robert McChesney—the history of the internet and its corporate owners. Rebuild Conference presentation. https://www.youtube.com/watch?v=bLCasOG3S2o

McComas, K., and J. Shanahan. 1999. Telling stories about global climate change measuring the impact of narratives on issue cycles. *Communication Research* 26 (1): 30–57.

McCright, A. M. 2006. Dealing with climate change contrarians. In *Creating a climate for change*, edited by S. Moser and L. Dilling, 200–212. Cambridge: Cambridge University Press.

McIntyre, Douglas. 2019. Over 2,000 American newspapers have closed in the past

15 years. *247wallst.com*, July 23. https://247wallst.com/media/2019/07/23/over-2000-american-newspapers-have-closed-in-past-15-years/

McQuail, D. 1977. The influences and effects of the mass media. In *Mass communication and society*, edited by James Curran, M. Gurevitch, and J. Woolocott, 70–94. London: Arnold.

Moser, S. C. 2010. Communicating climate change: History, challenges, process, and future directions. *Wiley Interdisciplinary Reviews: Climate Change* 1 (1): 31–53.

Nosty, B. D. 2014. Media and climate change: From content analysis to analysis content. The interdisciplinary construction of a new paradigm. *Interactions: Studies in Communication & Culture* 5 (1): 7–23.

Nuccitelli, Dana. 2015. Fox News' inner struggle with climate misinformation. *Guardian.com*, Aug 17. https://www.theguardian.com/environment/climate-consensus-97-per-cent/2015/aug/17/fox-news-inner-struggle-with-climate-misinformation

Olausson, U. 2009. Global warming—global responsibility? Media frames of collective action and scientific certainty. *Public Understanding of Science* 18 (4): 421–36.

Opel, A. 2006. Corporate culture keeps nature regular: The "super citizen," the media, and the "Metamucil and Old Faithful" ad. *Capitalism Nature Socialism* 17 (3): 100–113.

Park, D., V. Wang, and J. Pinto. 2015. Beyond nuclear disaster and risk: Comparison of post-Fukushima news in U.S. and German press. *Communication, Culture and Critique*, July 29. https://doi.org/10.1111/cccr.12119

Pew Research Center for People and the Press. 2006. Online papers modestly boost newspaper readership. http://pewresearch.org/politics/2006/07/30/online-papers-modestly-boot-newspaper-readership/

Philo, G., and C. Happer. 2013. *Communicating climate change and energy security: New methods in understanding audiences*. New York: Routledge.

Pope, K., and M. Hertsgaard. 2019. Why are the US news media so bad at covering climate change? *Guardian.com*, April 22. https://www.theguardian.com/environment/2019/apr/22/why-is-the-us-news-media-so-bad-at-covering-climate-change

Poynter. 2020. Here are the newsroom layoffs, furloughs and closures caused by the coronavirus. Poynter.org. https://www.poynter.org/business-work/2020/here-are-the-newsroom-layoffs-furloughs-and-closures-caused-by-the-coronavirus/

Public Citizen. 2018. Climate uncovered: Media fail to connect Hurricane Florence to climate change. Citizen.org. https://www.citizen.org/sites/default/files/public-citizen-climate-uncovered-media-fail-connect-hurricane-florence-climate-change-september-2018.pdf

Public Citizen. 2019. Major media fail to connect Hurricane Dorian to climate change. Public Citizen. https://www.citizen.org/news/major-media-fail-to-connect-hurricane-dorian-to-climate-change/

Ryghaug, M., K. H. Sørensen, and R. Næss. 2011. Making sense of global warming: Norwegians appropriating knowledge of anthropogenic climate change. *Public Understanding of Science* 20 (6): 778–95.

Schmidt, A., A. Ivanova, and M. S. Schäfer. 2013. Media attention for climate change around the world: A comparative analysis of newspaper coverage in 27 countries. *Global Environmental Change* 23 (5): 1233–48.

Schudson, Michael. 1989. Historical roots of consumer culture. In *Advertising in society: Classic and contemporary readings on advertising's role in society*, edited by Roxanne Hovland and Gary Wilcox, 42–72. Lincolnwood, IL: NTC Business Books.

Schuldt, J. P., and S. Roh. 2014. Media frames and cognitive accessibility: What do "global warming" and "climate change" evoke in partisan minds? *Environmental Communication* 8 (4): 529–48.

Shanahan, M. 2007. *Talking about a revolution: Climate change and the media.* London: International Institute for Environment and Development.

Simon, Mollie. 2019. An analysis of trends of print media coverage on climate change in the Trump era. University of Pennsylvania ScholarlyCommons. https://repository.upenn.edu/mes_capstones/79

Smith, J. 2005. Dangerous news: Media decision making about climate change risk. *Risk Analysis* 25 (6): 1471–82.

Sparks, Colin. 2018. Changing concepts for a changing world. *Journal of Communication* 68 (2): 390–98.

Stamm, K. R., F. Clark, and P. Reynolds-Eblacas. 2000. Mass communication and public understanding of environmental problems: The case of global warming. *Public Understanding of Science* 9 (3): 219–37.

Takahashi, B., K. Huang, F. Fico, and D. Poulson. 2017. Climate change reporting in Great Lakes region newspapers: A comparative study of the use of expert sources. *Environmental Communication* 11 (1): 106–21.

Tankard, J. W., and M. Ryan. 1974. News source perceptions of accuracy of science coverage. *Journalism and Mass Communication Quarterly* 51 (2): 219.

Trumbo, C. 1996. Constructing climate change: Claims and frames in US news coverage of an environmental issue. *Public Understanding of Science* 5 (3): 269–83.

Waterson, Jim. 2019. United States classed as "problematic" place for journalists to work. *Guardian.com*, April 18. https://www.theguardian.com/media/2019/apr/18/united-states-classed-as-problematic-place-for-journalists-to-work

Watts, N., M. Amann, N. Arnell, S. Ayeb-Karlsson, K. Belesova, H. Berry, and D. Campbell-Lendrum. 2018. The 2018 report of the Lancet Countdown on health and climate change: Shaping the health of nations for centuries to come. *The Lancet* 392 (10163): 2479–2514.

Weathers, M. R., and B. E. Kendall. 2015. Developments in the framing of climate change as a public health issue in US newspapers. *Environmental Communication* 10 (5): 593–611.

Weber, E., and P. Stern. 2011. Public understanding of climate change in the United States. *American Psychologist* 66 (4): 315–28.

"Whitening of the media": As pandemic cuts ravage newsrooms, diversity is the "first thing that goes." 2020. *Democracynow.org*, May 19. https://www.democracynow.org/2020/5/19/pandemic_media_cuts_journalism_jobs

Wilson, K. M. 1995. Mass media as sources of global warming knowledge. *Mass Communication Review* 22: 75–89.

Ytterstad, A. 2015. Framing global warming: Is that really the question? A realist, Gramscian critique of the framing paradigm in media and communication research. *Environmental Communication* 9 (1): 1–19.
Zehr, S. C. 2000. Public representations of scientific uncertainty about global climate change. *Public Understanding of Science* 9 (2): 85–103.

9. Misinformation, Blogs, and Public Opinion

Anderson, A. 2009. Media, politics and climate change: Towards a new research agenda. *Sociology Compass* 3 (2): 166–82.
Antilla, L. 2005. Climate of scepticism: US newspaper coverage of the science of climate change. *Global Environmental Change* 15 (4): 338–52.
Arlt, D., I. Hoppe, and J. Wolling. 2011. Climate change and media usage: Effects on problem awareness and behavioural intentions. *International Communication Gazette* 73 (1–2): 45–63.
Aronoff, Kate. 2019. Bernie Sanders's climate plan is more radical than his opponents'—and more likely to succeed. *Intercept.com*, August 22. https://theintercept.com/2019/08/22/bernie-sanders-climate-policy/
Block, B. 2010. Covering climate change. *World Watch* 23 (2). http://worldwatch.org/node/6373
Boussalis, C., and T. G. Coan. 2016. Text-mining the signals of climate change doubt. *Global Environmental Change* 36: 89–100.
Boykoff, M. T. 2013. Public enemy no. 1? Understanding media representations of outlier views on climate change. *American Behavioral Scientist* 57 (6): 796–817.
Boykoff, M. T., and J. M. Boykoff. 2004. Balance as bias: Global warming and the US prestige press. *Global Environmental Change* 14 (2): 125–36.
Boykoff, M. T., and J. Farrell. 2020. Climate change countermovement organizations and media attention in the United States. In *Climate change denial and public relations: Strategic communication and interest groups in climate inaction*, edited by Núria Almiron and Jordi Xifra, 121–39. New York: Routledge.
Boykoff, M. T., and J. T. Roberts. 2007. Media coverage of climate change: Current trends, strengths, weaknesses. *Human Development Report 2007/2008*, 1–53. New York: United Nations Development Programme.
Boykoff, M. T., and J. Smith. 2010. Media presentations of climate change. In *Routledge handbook of climate change and society*, edited by C. Lever-Tracy, 210–18. New York: Routledge.
Brulle, R. J., J. Carmichael, and J. C. Jenkins. 2012. Shifting public opinion on climate change: An empirical assessment of factors influencing concern over climate change in the US, 2002–2010. *Climatic Change* 114 (2): 169–88.
Carvalho, A. 2010. Media (ted) discourses and climate change: A focus on political subjectivity and (dis) engagement. *Wiley Interdisciplinary Reviews: Climate Change* 1 (2): 172–79.
Elgesem, D., L. Steskal, and N. Diakopoulos. 2015. Structure and content of the discourse on climate change in the blogosphere: The big picture. *Environmental Communication* 9 (2): 169–88.

Faber, Daniel. 2008. *Capitalizing on environmental injustice: The polluter-industrial complex in the age of globalization*. New York: Roman and Littlefield.

Fottrell, Quentin. 2018. 50 million American households can't even afford basic living expenses. *MarketWatch.com*, June 9. https://www.marketwatch.com/story/50-million-american-households-cant-afford-basic-living-expenses-2018-05-18

Goldenberg, S., and H. Bengtsson. 2015. Secretive donors gave US climate denial groups $125m over three years. *Guardian.com*, June 9. https://www.theguardian.com/environment/2015/jun/09/secretive-donors-gave-us-climate-denial-groups-125m-over-three-years

Good, J. E. 2008. The framing of climate change in Canadian, American, and international newspapers: A media propaganda model analysis. *Canadian Journal of Communication* 33 (2): 233.

Greitemeyer, T. 2013. Beware of climate change skeptic films. *Journal of Environmental Psychology* 35: 105–9.

Hackett, R. A. 2016. Dallas Smythe Award keynote lecture. *Democratic Communiqué* 27: 1–14.

Hackett, R. A. 2018. Planetary emergency and sustainable democracy: What can media and communication scholars do? *Political Economy of Communication* 6 (1): 98–106.

Hiles, S. S., and A. Hinnant. 2014. Climate change in the newsroom journalists' evolving standards of objectivity when covering global warming. *Science Communication*, May 19. https://doi.org/10.1177/1075547014534077

InfluenceMap. 2019. Big oil's real agenda on climate change. March. https://influencemap.org/report/How-Big-Oil-Continues-to-Oppose-the-Paris-Agreement-38212275958aa21196dae3b76220bddc

IPCC. 2014. *Mitigation of climate change*. Contribution of Working Group III to the Fifth Assessment Report of the Intergovernmental Panel on Climate Change. Edited by O. Edenhofer, R. Pichs-Madruga, Y. Sokona, E. Farahani, S. Kadner, K. Seyboth, A. Adler, I. Baum, S. Brunner, and P. Eickemeier, et al. Cambridge: Cambridge University Press.

Jacques, P. J., R. E. Dunlap, and M. Freeman. 2008. The organization of denial: Conservative think tanks and environmental scepticism. *Environmental Politics* 17 (3): 349–85.

Jurado, D. G., and C. J. Maldonado. 2014. Advertising and climate change, relationships and divergences. *Interactions: Studies in Communication & Culture* 5 (1): 107–25.

"Kochland": How David Koch helped build an empire to shape U.S. politics and thwart climate action. 2019. *Democracynow.org*, August 27. https://www.democracynow.org/2019/8/27/christopher_leonard_kochland_koch_brothers

Koteyko, N., R. Jaspal, and B. Nerlich. 2013. Climate change and "climategate" in online reader comments: A mixed methods study. *Geographical Journal* 179 (1): 74–86.

Leiserowitz, A., E. Maibach, S. Rosenthal, J. Kotcher, M. Ballew, M. Goldberg, and A. Gustafson. 2018. *Climate change in the American mind: December 2018*. Yale University and George Mason University. New Haven: Yale Program on Climate Change Communication.

Leiserowitz, A., E. Maibach, C. Roser-Renouf, G. Feinberg, and S. Rosenthal. 2016. Politics and global warming, Spring 2016. Yale University and George Mason University. New Haven: Yale Program on Climate Change Communication.

Lewis, J., and T. Boyce. 2009. Climate change and the media: The scale of the challenge. In *Climate change and the media*, edited by J. Lewis and T. Boyce, 3–16. New York: Peter Lang.

Lozada, Hector, Gary Kritz, and Alma Mintu-Wimsatt. 2017. Let's (not) talk about climate change: American consumers and global warming. Marketing Management Association Proceedings.

Matthews, P. 2015. Why are people skeptical about climate change? Some insights from blog comments. *Environmental Communication* 9 (2): 153–68.

Mazur, A., and J. Lee. 1993. Sounding the global alarm: Environmental issues in the US national news. *Social Studies of Science* 23 (4): 681–720.

McCright, A. M., and R. E. Dunlap. 2011. The politicization of climate change and polarization in the American public's views of global warming, 2001–2010. *Sociological Quarterly* 52 (2): 155–94.

Miller-Gaither, B., and T. K. Gaither. 2016. Marketplace advocacy by the US fossil fuel industries: Issues of representation and environmental discourse. *Mass Communication and Society* 19 (5): 585–603.

Milman, Oliver. 2020. Revealed: Quarter of all tweets about climate crisis produced by bots. *Guardian.com*, February 21. https://www.theguardian.com/technology/2020/feb/21/climate-tweets-twitter-bots-analysis

Milman, Oliver, and Fiona Harvey. 2019. US is hotbed of climate change denial, major global survey finds. *Guardian.com*, May 8. https://www.theguardian.com/environment/2019/may/07/us-hotbed-climate-change-denial-international-poll

Monmouth University Polling Institute. 2018. Climate concerns increase; most Republicans now acknowledge change. https://www.monmouth.edu/polling-institute/reports/monmouthpoll_us_112918

Nuccitelli, Dana. 2018. Republican lawmakers react to the IPCC report–"we have scientists" too! *Guardian.com*, October 17. https://www.theguardian.com/environment/climate-consensus-97-per-cent/2018/oct/17/republican-lawmakers-react-to-the-ipcc-report-we-have-scientists-too

Oreskes, N., and E. M. Conway. 2011. *Merchants of doubt: How a handful of scientists obscured the truth on issues from tobacco smoke to global warming*. New York: Bloomsbury.

Pearson, A. R., and J. P. Schuldt. 2015. Bridging climate communication divides beyond the partisan gap. *Science Communication* 37 (6): 805–12.

Pew Research Center. 2010. Bloggers focus on two favorite subjects: Health care and global warming. October 12–16. http://www.journalism.org/2010/04/08/bloggers-focus-two-favorite-subjects-health-care-and-global-warming/

Pew Research Center. 2016. Budget deficit slips as public priority. http://www.people-press.org/2016/01/22/budget-deficit-slips-as-public-priority/

Pew Research Center. 2017. The partisan divide on political values grows even wider. People-Press.org, October 5. https://www.people-press.org/2017/10/05/7-global-warming-and-environmental-regulation-personal-environmentalism/

Pidgeon, N. 2012. Climate change risk perception and communication: Addressing a critical moment? *Risk Analysis* 32 (6): 951–56.
Shanahan, M. 2007. *Talking about a revolution: Climate change and the media*. London: International Institute for Environment and Development.
Tabuchi, Hiroko. 2018. Spreading lies on climate science, and exploiting Google's algorithms to do it. *New York Times*, January 2, B3.
Van der Linden, S. L., A. A. Leiserowitz, G. D. Feinberg, and E.W. Maibach. 2015. The scientific consensus on climate change as a gateway belief: Experimental evidence. *PloS one* 10 (2): e0118489.
Watson, R., C. Carraro, P. Canziani, N. Nakicenovic, J. McCarthy, J. Goldemberg, and L. Hisas. 2016. The truth about climate change. Universal Ecological Fund. https://drive.google.com/drive/folders/0B7HRGKIJWol-OEZfNDFjRHUwM2c
Weathers, M. R., and B. E. Kendall. 2015. Developments in the framing of climate change as a public health issue in US newspapers. *Environmental Communication* 10 (5): 593–611.
Weber, E. and P. Stern. 2011. Public understanding of climate change in the United States. *American Psychologist* 66 (4): 315–28.
Whitmarsh, L. 2011. Scepticism and uncertainty about climate change: Dimensions, determinants and change over time. *Global Environmental Change* 21 (2): 690–700.

10. The Battle against Global Warming Is Also a Fight for Media Reform

Althaus, S., J. Edy, R. Entman, and P. Phalen. 1996. Revising the indexing hypothesis: Officials, media and the Libya crisis. *Political Communication* 13: 407–21. https://doi.org/10.1080/10584609.1996.9963128
Anderson, A. 2009. Media, politics and climate change: Towards a new research agenda. *Sociology Compass* 3 (2): 166–82.
Arlt, D., I. Hoppe, and J. Wolling. 2011. Climate change and media usage: Effects on problem awareness and behavioural intentions. *International Communication Gazette* 73 (1–2): 45–63.
Artz, Lee. 2012. 21st century socialism: Making a state for revolution. *tripleC* 10 (2): 537–54.
Baker, C. E. 1994. *Advertising and a democratic press*. Princeton: Princeton University Press.
Baran, P., and P. Sweezy. 2013. The quality of monopoly capitalist society: Culture and communications. *Monthly Review* 65 (3): 43–64.
Boykoff, M. T. 2007. From convergence to contention: United States mass media representations of anthropogenic climate change science. *Transactions of the Institute of British Geographers* 32 (4): 477–89.
Boykoff, M. T. 2013. Public enemy no. 1? Understanding media representations of outlier views on climate change. *American Behavioral Scientist* 57 (6): 796–817.
Brüggemann, M., and S. Engesser. 2014. Between consensus and denial: Climate journalists as interpretive community. *Science Communication* 36 (4): 399–427.
Carvalho, A. 2005. Representing the politics of the greenhouse effect: Discursive strategies in the British media. *Critical Discourse Studies* 2 (1): 1–29.

Carvalho, A. 2007. Ideological cultures and media discourses on scientific knowledge: Re-reading news on climate change. *Public Understanding of Science* 16 (2): 223–43.

Carvalho, A. 2010. Media (ted) discourses and climate change: A focus on political subjectivity and (dis) engagement. *Wiley Interdisciplinary Reviews: Climate Change* 1 (2): 172–79.

Chesebro, J., and D. T. McMahan. 2006. Media constructions of mass murder-suicides as drama: The *New York Times'* symbolic construction of mass murder-suicides. *Communication Quarterly* 54 (4): 407–25.

Dixit, K. 1994. The greening of the media. In *AMIC Conference on Communication, Convergence and Development: Bangkok, Jun 23–25, 1994*. Singapore: Asian Media Information and Communication Centre.

Emmons, A. 2016. Oil lobby paid *Washington Post* and *Atlantic* to host climate-change deniers at RNC. *Intercept.com*, July 22. https://theintercept.com/2016/07/22/oil-lobby-paid-washington-post-and-atlantic-to-host-climate-change-deniers-at-rnc/

Ereaut, G., and N. Segnit. 2006. Warm words: How we are telling the climate story and can we tell it better. Institute for Public Policy Research. https://www.ippr.org/publications/warm-wordshow-are-we-telling-the-climate-story-and-can-we-tell-it-better

Ettema. J. 2005. Crafting cultural resonance: Imaginative power in everyday journalism. *Journalism* 6 (2): 131–52.

Feldman, C., and E. Swanson. 2016. Poll: Getting facts right key to Americans' trust in media. Associated Press, April 17. http://bigstory.ap.org/article/35c595900e0a4ffd99fbdc48a336a6d8/poll-vast-majority-americans-dont-trust-news-media

Foster, J. B. 2017. *Trump in the White House: Tragedy and farce*. New York: Monthly Review Press.

Foster, J. B., and R. W. McChesney. 2013. The cultural apparatus of monopoly capital: An introduction. *Monthly Review* 65 (3): 1–33.

Fuchs, C. 2015. *Culture and economy in the age of social media*. New York: Routledge.

Gibson, T. A., R. T. Craig, A. C. Harper, and J. M. Alpert. 2015. Covering global warming in dubious times: Environmental reporters in the new media ecosystem. *Journalism*, January 4. https://doi.org/10.1177/1464884914564845

Gilens, M., and C. Hertzman. 2000. Corporate ownership and news bias: Newspaper coverage of the 1996 Telecommunications Act. *Journal of Politics* 62 (2): 369–86.

Good, J. E. 2008. The framing of climate change in Canadian, American, and international newspapers: A media propaganda model analysis. *Canadian Journal of Communication* 33 (2): 233.

Hackett, R. A. 2016. Dallas Smythe Award keynote lecture. *Democratic Communiqué* 27: 1–14.

Häussler, T., H. Schmid-Petri, S. Adam, U. Reber, and D. Arlt. 2016. The climate of debate: How institutional factors shape legislative discourses on climate change. A comparative framing perspective. *Studies in Communication Sciences* 16: 94–102.

Herman, E. S., and N. Chomsky. 2010. *Manufacturing consent: The political economy of the mass media*. New York: Pantheon Books.

Hestres, L. E. 2013. Preaching to the choir: Internet-mediated advocacy, issue public mobilization, and climate change. *New Media & Society* 16 (2): 323–39.

Jameson, F. 1994. Postmodernism and the market. In *Mapping Ideology*, edited by S. Žižek, 278–95. London, Verso.

Kinsella, W. J. 2014. Rearticulating nuclear power: Energy activism and contested common sense. *Environmental Communication* 9 (3). https://doi.org/10.1080/17524032.2014.978348

Kitch, C., and J. Hume. 2008. *Journalism in a culture of grief*. New York: Routledge.

Lewis, J., and T. Boyce. 2009. Climate change and the media: The scale of the challenge. In *Climate change and the media*, edited by J. Lewis and T. Boyce, 3–16. New York: Peter Lang.

McChesney, R. W. 2013. *Digital disconnect: How capitalism is turning the internet against democracy*. New York: New Press.

McCright, A. M. 2006. Dealing with climate change contrarians. In *Creating a climate for change*, edited by S. Moser and L. Dilling, 200–212. Cambridge: Cambridge University Press.

McCright, A. M., and R. E. Dunlap. 2003. Defeating Kyoto: The conservative movement's impact on US climate change policy. *Social Problems* 50 (3): 348–73.

Milman, Oliver, and Fiona Harvey. 2019. US is hotbed of climate change denial, major global survey finds. *The Guardian*, May 8. https://www.theguardian.com/environment/2019/may/07/us-hotbed-climate-change-denial-international-poll

Morgan, T. 2015. Growing ourselves to death? Economic and ecological crises, the growth of waste, and the role of the media and cultural industries. *Human Geography* 8 (1): 68–81.

Mueller, Rowitha. 1989. *Bertolt Brecht and the theory of the media*. Lincoln: University of Nebraska Press.

Nielsen Company. 2007. Global Omnibus Survey. Oxford, UK.

Nyhan, B. 2014. Scandal potential: How political context and news congestion affect the president's vulnerability to media scandal. *British Journal of Political Science* 45 (2): 1–32. https://doi.org/10.1017/S0007123413000458

Olausson, U. 2009. Global warming—global responsibility? Media frames of collective action and scientific certainty. *Public Understanding of Science* 18 (4): 421–36.

Olausson, U. 2010. Towards a European identity? The news media and the case of climate change. *European Journal of Communication* 25 (2): 138–52.

Oltermann, P. 2016. German coalition agrees to cut carbon emissions up to 95% by 2050. *The Guardian*, November 11. https://www.theguardian.com/environment/2016/nov/11/german-coalition-agrees-to-cut-carbon-emissions-up-to-95-by-2050

Park, D., A. Tajima, S. Nah, and S. Nichols. 2004. Party politics and public perspectives: Pre-invasion media coverage and political agendas. *Global Media Journal* 3 (5).

Park, D., V. Wang, and J. Pinto. 2015. Beyond nuclear disaster and risk: Comparison of post-Fukushima news in U.S. and German press. *Communication, Culture and Critique*, July 29. https://doi.org/10.1111/cccr.12119

Park, D., and L. Wright. 2007. Giving labor the business? Business and labor reporting from 1980–2000. *Labor Studies Journal* 32 (1): 60–81.

Perry, R. 2015. Reviving the "liberal media" myth. *Consortiumnews.org*, November 2. https://consortiumnews.com/2015/11/02/reviving-the-liberal-media-myth/

Peters, H. P., and H. Heinrichs. 2008. Legitimizing climate policy: The "risk construct" of global climate change in the German mass media. *International Journal of Sustainability Communication* 3 (1): 14–36.

Pew Research Center. 2009. Press accuracy rating hits two decade low. http://www.people-press.org/2009/09/13/press-accuracy-rating-hits-two-decade-low/

Pew Research Center. 2013. Public esteem for military still high. http://www.pewforum.org/2013/07/11/public-esteem-for-military-still-high/#about

Pickard, V. 2019. *Democracy without journalism? Confronting the misinformation society*. Oxford: Oxford University Press.

Post, S., and N. Ramirez. 2018. Politicized science communication: Predicting scientists' acceptance of overstatements by their knowledge certainty, media perceptions, and presumed media effects. *Journalism & Mass Communication Quarterly* 95 (4): 1150–70.

Reich, Robert. 2019. Facts are under siege: Now more than ever, we need to invest in journalism. *Guardian.com*, December 2. https://www.theguardian.com/commentisfree/2019/dec/02/facts-are-under-siege-now-more-than-ever-we-need-to-invest-in-journalism

Reporters without Borders. 2016. 2016 World press freedom index. Reporters without Borders. https://rsf.org/en/ranking#

Roscho, B. 1975. *Newsmaking*. Chicago: University of Chicago Press.

Schoenfeld, A. C., R. F. Meier, and R. J. Griffin. 1979. Constructing a social problem: The press and the environment. *Social Problems* 27 (1): 38–61.

Schutz, E. 2011. *Inequality and power: The economics of class*. New York: Routledge.

Shanahan, M. 2007. *Talking about a revolution: Climate change and the media*. London: International Institute for Environment and Development.

Sparks, Colin. 2018. Changing concepts for a changing world. *Journal of Communication* 68 (2): 390–98.

Splichal, Slavko, and Boris Mance. 2018. Paradigm(s) lost? Islands of critical media research in communication journals. *Journal of Communication* 68 (2): 399–414.

Takahashi, B. 2011. Framing and sources: A study of mass media coverage of climate change in Peru during the V ALCUE. *Public Understanding of Science* 20 (4): 543–57.

Tuchman, G. 1978. *Making news: A study in the construction of reality*. New York: Free Press.

Van Dijk, T. A. 2013. *News analysis: Case studies of international and national news in the press*. Hillsdale, NJ: Erlbaum.

Watson, B. R. 2014. When critical voices should speak up: Patterns in news coverage of unofficial sources during the BP oil spill. *Journalism Practice* 8 (6): 842–54. https://doi.org/10.1080/17512786.2014.892723

Weber, E., and P. Stern. 2011. Public understanding of climate change in the United States. *American Psychologist* 66 (4): 315–28.

Weingart, P., A. Engels, and P. Pansegrau. 2000. Risks of communication: Discourses on climate change in science, politics, and the mass media. *Public Understanding of Science* 9 (3): 261–83.

Conclusion

Hertsgaard, Mark. 2019. "We're losing the race": UN secretary general calls climate change an "emergency." *Guardian.com*, September 18. https://www.theguardian.com/environment/2019/sep/18/un-secretary-general-climate-crisis-trump

Hestres, Luis. 2020. Fighting climate change denial in the United States. In *Climate change denial and public relations: Strategic communication and interest groups in climate inaction*, edited by Núria Almiron and Jordi Xifra, 217–32. New York: Routledge.

Holden, Emily, and Laurent Gambino. 2019. Sanders to unveil $16tn climate plan, far more aggressive than rivals' proposals. *Guardian.com*, August 22. https://www.theguardian.com/us-news/2019/aug/22/bernie-sanders-climate-change-plan

Karlin, Sam. 2019. Where do Louisiana governor candidates stand on climate change? *Advocate.com*, September 8. https://www.theadvocate.com/baton_rouge/news/politics/elections/article_29b54b52-d0cc-11e9-a12a-b30a82005885.htmlhttps://www.theadvocate.com/baton_rouge/news/politics/elections/article_29b54b52-d0cc-11e9-a12a-b30a82005885.html

Krishnaswamy, Esha. 2019. Climate minimizers don't deny climate change—but find endless reasons to reject Sanders' plan to stop it. *Fair.org*, September 6. https://fair.org/home/climate-minimizers-dont-deny-climate-change-but-find-endless-reasons-to-reject-sanders-plan-to-stop-it/?awt_l=Ivb6m&awt_m=it4N8eHev0OI_TQ

Laville, Sandra. 2019. Swedish newspaper stops taking adverts from fossil fuel firms. *Guardian.com*, September 26. https://www.theguardian.com/environment/2019/sep/26/swedish-newspaper-stops-taking-adverts-from-fossil-fuel-firms

Lewis, J., and T. Boyce. 2009. Climate change and the media: The scale of the challenge. In *Climate change and the media*, edited by J. Lewis and T. Boyce, 3–16. New York: Peter Lang.

Perry, Catie. 2019. *New York Times* drops oil sponsorship and sparks debate: Should news orgs back agenda specific events? *Foxbusiness.com*, September 3. https://www.foxbusiness.com/media/new-york-times-sponsorship-oil-and-money-conference

Pickard, V. 2019. *Democracy without journalism? Confronting the misinformation society.* Oxford: Oxford University Press.

Waterson, Jim. 2020. *Guardian* to ban advertising from fossil fuel firms. *Guardian.com*, January 29. https://www.theguardian.com/media/2020/jan/29/guardian-to-ban-advertising-from-fossil-fuel-firms-climate-crisis

White, Damian. 2019. Just transitions/design for transitions: Preliminary notes on a design politics for a Green New Deal. *Capitalism Nature Socialism* 31 (2). https://doi.org/10.1080/10455752.2019.1583762

Index